Vectors and Transformations in Plane Geometry

Vectors and Transformations in Plane Geometry

Philippe Tondeur

University of Illinois at Urbana-Champaign

PUBLISH OR PERISH, INC. 1993

Publish or Perish, Inc.
Houston, TX

PREFACE

The first goal of this book is to explain the geometry of the plane by vector methods, in contrast to a synthetic approach. The vector approach is simple and direct. It represents a general method, while the synthetic approach has to many students the aspect of consisting of a multitude of flashes of insight. Each approach has its own charm.

The second goal is to introduce the student to the concept of transformation. This approach to geometry dates back to Felix Klein (1849–1925) and Sophus Lie (1842–1899). It gives concrete examples leading to an appreciation of the theory of groups, but does not require any previous knowledge of group theory.

Here is a summary of the contents. In Chapter 1 it is explained how one can do plane geometry by using vector methods. The axioms of geometry are embodied in the rules of calculations with vectors. The chapter begins with those rules, and ends with some geometrical facts proved via vector methods. The dominant idea in Chapter 2 is to look at all this from the point of view of groups of transformations. The group of dilatations is the group most appropriate for the choice of topics in Chapter 1. The discussion of the abstract group concept is followed by the examples of the dihedral groups, the symmetry groups of regular polygons. In Chapter 3 the notions of length and angle measurements are discussed in terms of a scalar product. In Chapter 4 we turn to the discussion of the group of isometries. The chapter ends with a classification of all finite groups of isometries. In Chapter 5 the link is made between linear maps and their representations by matrices. This is followed by solutions to the odd-numbered exercises.

The text has as prerequisites only high-school geometry and algebra. The typical student in this course at the University of Illinois is currently a Junior, frequently taking simultaneously a standard course in abstract algebra. For many undergraduate mathematics majors, this is the only geometry course they will take. The material can be covered comfortably in one semester. There are exercises throughout the text.

I am indebted to Mary-Elisabeth Hamstrom for many helpful comments on this text. Hilda Britt's unwavering word processing over the years has greatly simplified my life. Thanks also to several classes of students of Math 303 at the University of Illinois for their interest in the preliminary versions of this text. Some funny aspects of my immigrant English might remain. I hope that it will amuse some readers and irritate few.

Philippe Tondeur

CONTENTS

Chapter 4   Isometries

Chapter 5   Linear Maps and Matrices

Vectors and Transformations in Plane Geometry

CHAPTER 1 VECTORS IN THE PLANE

In this chapter we define vectors in the plane and discuss the rules of calculations. This is used to prove some simple geometric facts about planar figures. The axioms of geometry are embedded in the rules of calculations with vectors, and this chapter begins with these rules. All the properties considered in this chapter concern "affine geometry". The meaning of this concept will become clearer as we go through Chapter 1. The main point of this vector approach to geometry in contrast to the "analytic geometry" approach via coordinates is its simplicity when we try to prove facts about plane geometry.

1.1 Definitions. The position of a point A in the plane is characterized by two numbers (a_1, a_2), its Cartesian coordinates with respect to a pair of orthogonal (perpendicular) lines. If $A = (a_1, a_2)$ and $B = (b_1, b_2)$ are two points, the vector $\overrightarrow{AB}$ is given by

$$\overrightarrow{AB} = (b_1 - a_1, b_2 - a_2). \tag{1}$$

We visualize this as an arrow from A (the beginning point) to B (the end point).

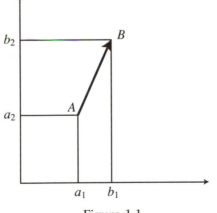

Figure 1.1.

See Figure 1.1. The vector $\overrightarrow{AB}$ is said to be located at A. Thus for $A = (a_1, a_2)$,

1

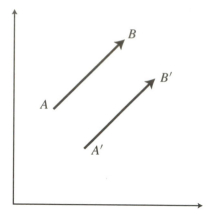

Figure 1.2.

$A' = (a_1', a_2')$ and $B = (b_1, b_2)$, $B' = (b_1', b_2')$ we define equality by

$$\overrightarrow{AB} = \overrightarrow{A'B'} \quad \text{if and only if} \quad b_1 - a_1 = b_1' - a_1', \quad b_2 - a_2 = b_2' - a_2'. \qquad (2)$$

Note that the beginning points A and A' need not be the same. The intuitive meaning of $\overrightarrow{AB} = \overrightarrow{A'B'}$ is that these vectors are parallel and of the same magnitude or length. Their direction is the same. See Figure 1.2.

The point $O = (0,0)$ is the intersection of the two orthogonal coordinate lines. The point A can be identified with the vector $\overrightarrow{OA}$, since both are given by (a_1, a_2). Thus one can write A instead of $\overrightarrow{OA}$. O is called the zero vector or origin.

1.2 Addition of vectors. Let $A = (a_1, a_2)$ and $B = (b_1, b_2)$ be two vectors. One defines a new vector $A + B$ by

$$A + B = (a_1 + b_1, a_2 + b_2). \qquad (3)$$

This corresponds to the parallelogram construction in Figure 1.3. This operation of addition of vectors has the following properties:

(A1)	$A + B = B + A$	commutativity
(A2)	$A + (B + C) = (A + B) + C$	associativity
(A3)	$O + A = A + O = A$	identity
(A4)	$A + (-A) = O$	inverse.

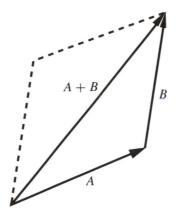

Figure 1.3.

In the last property, the vector $-A$ is given by $-A = (-a_1, -a_2)$. These properties derive immediately from the definition (2). A simplifying definition is the convention

$$A - B = A + (-B). \tag{4}$$

$A - B$ is the vector one adds to B to get A. This is illustrated in Figure 1.4. If

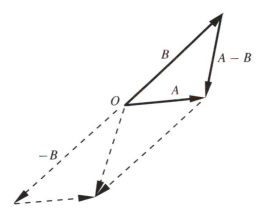

Figure 1.4.

vectors are identified with points, then

$$A - B = \overrightarrow{BA}. \tag{5}$$

In particular

$$A - O = \overrightarrow{OA} = A. \tag{6}$$

Figure 1.5.

1.3 Multiplication by a scalar. Let $A = (a_1, a_2)$ and r be a real number (a scalar). One defines a new vector rA by

$$rA = (ra_1, ra_2). \qquad (7)$$

In Figure 1.5, $r = \frac{1}{2}$. Multiplying by r rescales the vector A. If $r > 1$, rA is a longer vector. If $0 < r < 1$, rA is a shorter vector. If $r = 1$, the result is A itself, i.e., $1A = A$. If $r = 0$, the result is the zero vector, i.e., $0A = O$—note that in this equation the 0 on the left is a number, while the O on the right denotes the zero vector. If A is not the zero vector, the scalar multiples rA all lie on a line through the origin and the point at the endpoint of the arrow representing A. This we call the **line along** A or the **line through the points** O **and** A. We also say that A is a vector along this line. For $r > 0$, we get the points on the ray (half-line) from O through A, while for $r < 0$, we get the points on the opposite ray. In particular for $r = -1$ we get

$$(-1)A = (-1)(a_1, a_2) = (-a_1, -a_2) = -A. \qquad (8)$$

Multiplication by a scalar has the following properties:

(M1) $\qquad\qquad (r + s)A = rA + sA$ $\left.\vphantom{\begin{array}{c}a\\a\end{array}}\right\}$ distributivity

(M2) $\qquad\qquad r(A + B) = rA + rB$

(M3) $\qquad\qquad r(sA) = (rs)A$

(M4) $\qquad\qquad 1A = A.$

These properties derive from (3) and (7).

EXERCISE 1.1. Write out the proof of (M2).

Two particularly important vectors are $E_1 = (1, 0)$ and $E_2 = (0, 1)$, called the **standard basis vectors**. The scalar multiples $x_1 E_1 = (x_1, 0)$ give the points of the line ℓ_{OE_1}, the line through O and E_1 (the first coordinate axis), and similarly the scalar multiples $x_2 E_2 = (0, x_2)$ give the points of the second coordinate axis ℓ_{OE_2}. Since $A = (a_1, a_2)$ can be written

$$A = (a_1, 0) + (0, a_2) = a_1 E_1 + a_2 E_2,$$

this says that every vector A has a unique representation as a sum of one vector along the first coordinate axis and one vector along the second coordinate axis. This expresses A as the fourth vertex of a rectangle whose other vertices are O, $(a_1, 0)$ and $(0, a_2)$.

Consider more generally two nonzero vectors B, C such that the line ℓ_{OB} through O and B and the line ℓ_{OC} through O and C are distinct. Then every vector A has a unique representation

$$A = x_1 B + x_2 C \tag{D}.$$

To prove this for $A = (a_1, a_2)$ and $B = (b_1, b_2)$, $C = (c_1, c_2)$ amounts to proving that the system of linear equations

$$a_1 = x_1 b_1 + x_2 c_1$$
$$a_2 = x_1 b_2 + x_2 c_2$$

has a unique solution x_1, x_2. The necessary and sufficient condition for this to be true is that $b_1 c_2 - b_2 c_1 \neq 0$. This is precisely the condition that the lines ℓ_{OB} and ℓ_{OC} are distinct. To see this, note that the condition $b_1 c_2 - b_2 c_1 = 0$ implies $b_1/b_2 = c_1/c_2$ or $b_2/b_1 = c_2/c_1$, hence $B = \lambda C$ or $C = \mu B$.

The existence of distinct lines together with property (D) express the 2-dimensionality of the plane. We will use this as the definition of its 2-dimensionality.

1.4 Formal calculations. The rules (A1) to (A4) and (M1) to (M4) are simple and easily proved. The main point is that they themselves suffice to do all formal manipulations of vector equations. There is no need to go back to the explicit definition of the vectors involved. In this sense, these rules embody the essence of vectors, or what is the same, they form a system of axioms for vectors. This is illustrated in the following exercises. The assertions are to be proved using only the rules (A1) to (A4) and (M1) to (M4) and results of previous exercises.

EXERCISE 1.2. Prove that $A + B = A + C$ implies $B = C$.

EXERCISE 1.3. Prove that $rO = O$.

EXERCISE 1.4. Prove that $0A = O$.

EXERCISE 1.5. Prove that $rA = O$ implies that either $r = 0$ or $A = O$.

Together with the dimension property (D) above, the rules (A1) to (A4) and (M1) to (M4) characterize the plane. Informally, "affine plane geometry" is the geometry we can do with these rules alone. This is in contrast with "Euclidean plane geometry," which in addition involves length and angle measurements, as discussed fully in Chapter 3.

1.5 Equation of a line. Let A and B be two distinct points, and P a point of the line ℓ_{AB} determined by A and B. There is then a unique scalar t such that

$$P - A = t(B - A) \tag{9}$$

or equivalently, $\overrightarrow{AP} = t\overrightarrow{AB}$. To indicate the association of t with P, one writes

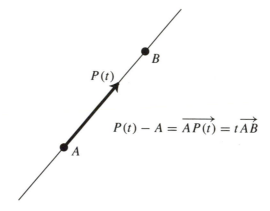

Figure 1.6.

$P = P(t)$. In Figure 1.6, the number t is approximately $3/5$. For $t = 1$, one has $P(1) = B$. For $t = 0$, one has $P(0) - A = O$, hence $P(0) = A$. For $0 < t < 1$, $P(t)$ is a point of the line segment on ℓ_{AB} between A and B. For $t > 1$ and for $t < 0$ the point $P(t)$ is outside this line segment. The situation is completely described as follows.

THEOREM 1.1. *Let A and B be two distinct points of the plane. Then the points P of the line ℓ_{AB} determined by A and B are precisely the vectors of the form*

$$P = aA + bB \quad \text{with} \quad a + b = 1. \tag{10}$$

Proof: If $P \in \ell_{AB}$, then by (9)

$$P = A + t(B - A) = (1 - t)A + tB.$$

If $a = 1 - t$ and $b = t$, this shows the existence of a representation of P as in (10). Conversely, a vector P as in (10) with $a + b = 1$ is of the form

$$P = aA + bB = (1 - b)A + bB.$$

If $t = b$, this equation implies $P - A = t(B - A)$, which is (9) and implies $P \in \ell_{AB}$.

This leads to the uniqueness question for a and b representing a given vector $P \in \ell_{AB}$ as in (10).

PROPOSITION 1.2. *Let A and B be two distinct points of the plane. If*

$$aA + bB = a'A + b'B \tag{11}$$

for pairs of numbers satisfying $a + b = 1$, $a' + b' = 1$, then $a = a'$ and $b = b'$.

Proof: The assumption (11) can be rewritten in the form

$$(a - a')A + (b - b')B = O. \tag{12}$$

But $(a - a') + (b - b') = (a + b) - (a' + b') = 0$. Thus if we denote $a - a'$ by r, then $b - b' = -r$. Therefore (12) implies $r(A - B) = rA - rB = O$. This in turn implies either $r = 0$ or $A - B = O$. But A and B are distinct by assumption. Thus $r = 0$ or $a = a'$. Therefore also $b = b'$.

EXERCISE 1.6. The *midpoint M* of points A and B or of the segment AB is given by the equation $\overrightarrow{AM} = \overrightarrow{MB}$ or $M - A = B - M$. Prove that $M = \frac{1}{2}(A + B)$. Do A and B have to be distinct?

We can now discuss the concept of parallel lines. If the lines ℓ_{AB} and ℓ_{CD} are defined by pairs of distinct points A, B, and C, D respectively, then ℓ_{AB} and ℓ_{CD} are *parallel*, $\ell_{AB} \mathbin{/\!/} \ell_{CD}$, if for some t we have $D - C = t(B - A)$ (or equivalently, if for some s we have $B - A = s(D - C)$). Note that necessarily $t \neq 0$. If $t < 0$, this means that the orientations of ℓ_{AB} and ℓ_{CD} are opposite. With this definition, one can now show that parallel lines ℓ_{AB} and ℓ_{CD} with C not on ℓ_{AB} have no common point. Prove this using Theorem 1.1. After these remarks, we turn to a discussion of parallelograms.

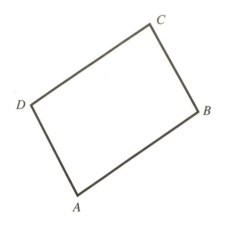

Figure 1.7.

1.6 Parallelograms. Four points A, B, C, D (in this order) define a quadrilateral $ABCD$ if A is joined to B, B to C, C to D, and finally D to A. When is quadrilateral $ABCD$ a parallelogram? A necessary condition is that $\overrightarrow{AB} = \overrightarrow{DC}$, or in equivalent notation $B - A = C - D$. Also equivalent is the identity $A + C = B + D$ (first + third = second + fourth). We adopt this as the defining property.

DEFINITION. The points A, B, C, D (in this order) define a parallelogram $ABCD$ provided that

$$A + C = B + D. \tag{13}$$

Note that this identity is not only equivalent to $B - A = C - D$, but also to $D - A = C - B$.

If A, B, C, D define a parallelogram, so do B, C, D, A (why?). The defining equation (13) multiplied by $\frac{1}{2}$ yields

$$\frac{1}{2}(A + C) = \frac{1}{2}(B + D). \tag{14}$$

This proves the following result (recall the definition of midpoints given in Exercise 1.6.)

PROPOSITION 1.3. *A quadrilateral is a parallelogram if and only if the diagonals bisect each other.*

In this discussion no assumption was made about the distinctness of A, B, C, D. Certainly the formula (13) still makes sense if some of the points coincide. To see if there is substance to this generalized point of view, consider special cases. E.g., $A = D$ implies $B = C$, which can be thought of as a degenerate case of a parallelogram.

EXERCISE 1.7. Let A, B, C, D be arbitrary points. Define M_1 as the midpoint of A and B, M_2 as the midpoint of B and C, M_3 as the midpoint of C and D, and M_4 as the midpoint of D and A.

(a) Prove that M_1, M_2, M_3, M_4 define a parallelogram.

(b) Discuss the extreme cases $A = B$ and $A = B = C$ (is the statement of (a) still meaningful in these cases?).

EXERCISE 1.8. (a) Consider a triangle, $\triangle ABC$, and the midpoints of the edges, $A' = \frac{1}{2}(B + C)$, $B' = \frac{1}{2}(A + C)$, $C' = \frac{1}{2}(A + B)$. Then the line $\ell_{A'B'}$ is parallel to the line ℓ_{AB}. Similarly $\ell_{A'C'}$ is parallel to ℓ_{AC}, and $\ell_{B'C'}$ is parallel to ℓ_{BC}. Verify the formula $A' - B' = C' - A = \frac{1}{2}(B - A)$.

(b) Let A', B', C' be three points which are not collinear. Find all triangles $\triangle ABC$ with $\triangle A'B'C'$ as the triangle of midpoints of its sides.

EXERCISE 1.9. Let the situation be as in Exercise 1.7. Consider the same construction carried out also for the points A, C, B, D (watch the order). This results in a second parallelogram $N_1 N_2 N_3 N_4$. Note that $N_2 = M_2$ and $N_4 = M_4$, so that the two resulting parallelograms have one diagonal in common. What can one conclude from this? Draw a picture.

1.7 Centroid of a triangle. Consider a triangle $\triangle ABC$. A median is a line joining a vertex to the midpoint of the opposite side (edge), such as $\ell_{CC'}$ in Figure 1.8.

THEOREM 1.4. *The medians of a triangle are* **concurrent**, *i.e., they intersect in one point (called the* **centroid** *of the triangle). This is illustrated in Figure 1.9.*

Proof: The midpoint of segment AB is given by $C' = \frac{1}{2}(A + B)$. A point of the median $\ell_{CC'}$ is of the form

$$P(t) = (1 - t)C + tC' = (1 - t)C + \frac{t}{2}(A + B).$$

Similarly a point of the median $\ell_{BB'}$ is of the form

$$Q(s) = (1 - s)B + \frac{s}{2}(A + C).$$

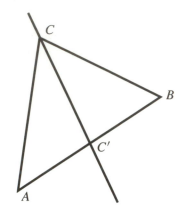

Figure 1.8.

Observe that $P(t) = Q(s)$ if $\dfrac{t}{2} = \dfrac{s}{2}$ (the coefficients of A) and $\dfrac{t}{2} = 1 - s$ (the coefficients of B). It follows that $t = s = \frac{2}{3}$, and the corresponding point

$$\frac{1}{3}(A + B + C)$$

is common to these two medians. The symmetry of this expression guarantees that this must be a point common to all medians.

The proof moreover gives the formula

$$G = \frac{1}{3}(A + B + C) \tag{15}$$

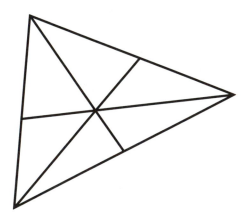

Figure 1.9.

for the centroid of $\triangle ABC$.

EXERCISE 1.10. Assuming formula (15) for the centroid, show that t as in the formula above for $P(t)$ has to equal $\frac{2}{3}$.

EXERCISE 1.11. Prove that the centroid of a triangle is also the centroid of the triangle of the midpoints of its sides.

EXERCISE 1.12. Consider Figure 1.10. (a) State and (b) prove the theorem suggested by this figure.

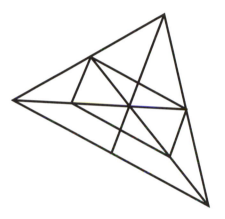

Figure 1.10.

EXERCISE 1.13. Let $ABCD$ be a parallelogram. Prove that the lines joining A to the midpoints of the opposite sides trisect the diagonal not through A.

1.8 Centroid of a finite point set. We generalize the preceding discussion to an arbitrary number of points. Let $A_1, \ldots, A_n$ be n points in the plane. The centroid of this set is defined by

$$G = \frac{1}{n}(A_1 + A_2 + \cdots + A_n). \tag{16}$$

For $n = 2$ this is the midpoint of the segment $A_1 A_2$, for $n = 3$ it is the centroid of a triangle. The following statement generalizes Theorem 1.4 to the case $n = 4$.

THEOREM 1.5. *The four lines through the points A_i $(i = 1, \ldots, 4)$ and the centroid of the remaining three points intersect in the centroid of A_1, A_2, A_3, A_4.*

Proof: Generalizing the argument in the proof of Theorem 1.4, we consider the point

$$\hat{G}_4 = \frac{1}{3}(A_1 + A_2 + A_3). \qquad (A_4 \text{ omitted})$$

We want to show that $G = \frac{1}{4}(A_1 + A_2 + A_3 + A_4)$ is on the line $\ell_{\hat{G}_4 A_4}$, i.e., for some t

$$G = \hat{G}_4 + t(A_4 - \hat{G}_4) = (1-t)\hat{G}_4 + tA_4 = \frac{1-t}{3}(A_1 + A_2 + A_3) + tA_4.$$

A sufficient condition for this is clearly $t = \frac{1}{4}$, and indeed, for $t = \frac{1}{4}$ this identity holds. By reasons of symmetry, the same argument proves that G must also lie on the corresponding lines $\ell_{\hat{G}_1 A_1}$, $\ell_{\hat{G}_2 A_2}$, $\ell_{\hat{G}_3 A_3}$ and hence is their common intersection point.

What is the situation if we join midpoints?

PROPOSITION 1.6. *The centroid G of A_1, A_2, A_3, A_4 is also on the line joining the midpoints of A_1, A_2 and A_3, A_4; in fact it is their midpoint.*

Proof: If $M = \frac{1}{2}(A_1 + A_2)$ and $N = \frac{1}{2}(A_3 + A_4)$, then

$$\frac{1}{2}(M + N) = \frac{1}{2}\left[\frac{1}{2}(A_1 + A_2) + \frac{1}{2}(A_3 + A_4)\right] = G.$$

EXERCISE 1.14. (a) What is the statement corresponding to Theorem 1.5 for arbitrary n? (b) Prove the statement.

A statement generalizing Theorem 1.5 and Proposition 1.6 is as follows. Let

$$A_1, \ldots, A_p; \ A_{p+1}, \ldots, A_{p+q}$$

be $n = p + q$ points, grouped into subsets of p and q points as indicated by the subscripts. Let U be the centroid of the first p points, and V the centroid of the remaining q points.

PROPOSITION 1.7. *In this situation, the centroid G of $A_1, \ldots, A_n$ lies on the line ℓ_{UV}; in fact,*

$$G = \frac{1}{n}(p \cdot U + q \cdot V).$$

Proof: A point of the line ℓ_{UV} has a unique representation

$$P(t) = U + t(V - U) = (1 - t)U + tV, \tag{17}$$

where

$$U = \frac{1}{p}(A_1 + \cdots + A_p), \quad V = \frac{1}{q}(A_{p+1} + \cdots + A_{p+q}).$$

A sufficient condition for G to be represented in this fashion is obviously $t = \dfrac{q}{n}$, and for $t = \dfrac{q}{n}$ the right-hand side of (17) is indeed G, since then

$$(1 - t)\frac{1}{p} = \left(1 - \frac{q}{n}\right) \cdot \frac{1}{p} = \frac{n - q}{n} \cdot \frac{1}{p} = \frac{1}{n}.$$

EXERCISE 1.15. Let $ABCD$ be a quadrilateral.

(a) There are six possible midpoints of pairs of points. To each such midpoint corresponds a unique other midpoint (of the remaining two points), to which it can be joined. This gives three lines, which have a common intersection. Where? (See also Exercise 1.9.) It may be helpful to visualize the situation spatially and think of the points A, B, C, D as the vertices of a tetrahedron.

(b) Prove the identity

$$\frac{1}{2}(B + C) - \frac{1}{2}(A + B) = \frac{1}{2}(C - A) = \frac{1}{2}(C + D) - \frac{1}{2}(D + A)$$

and explain its geometric meaning.

1.9 Centroid of the zeros of a complex polynomial. We wish to apply these ideas to the set of zeros of a polynomial with complex coefficients. For this purpose we use complex numbers. The idea is to write points of the plane in the form

$$z = x + iy$$

(instead of $P = (x, y)$). The addition of vectors corresponds then to the following addition operation for $z = x + iy$, $z' = x' + iy'$:

$$z + z' = (x + x') + i(y + y').$$

The rules (A1) to (A4) of Section 1.2 certainly hold. The operation

$$rz = rx + iry$$

for a real number r corresponds to the multiplication of a vector by a scalar. The rules (M1) to (M4) of Section 1.3 hold.

This viewpoint allows us to define the further multiplication operation

$$z \cdot z' = (xx' - yy') + i(xy' + yx'),$$

for complex numbers $z = x + iy$, $z' = x' + iy'$. Formally these expressions are multiplied by observing the distributive law and postulating $i^2 = -1$ for the square of the imaginary unit i. This operation is then associative and commutative on the set of complex numbers. The **complex conjugate** of a complex number $z = x + iy$ is the complex number

$$\bar{z} = x - iy.$$

Note that

$$z \cdot \bar{z} = x^2 + y^2,$$

which is denoted $|z|^2$. The reciprocal of a complex number is given by

$$\frac{1}{z} = \frac{1}{z} \cdot \frac{\bar{z}}{\bar{z}} = \frac{1}{|z|^2}\bar{z}.$$

Now one can perform all the usual rational operations with complex numbers. For instance, for $z_1 = x_1 + iy_1$, $z_2 = x_2 + iy_2 \neq 0$ the quotient is defined by

$$\frac{z_1}{z_2} = z_1 \cdot \frac{1}{z_2} = \frac{1}{|z_2|^2} \cdot z_1 \bar{z}_2.$$

The quadratic equation

$$az^2 + bz + c = 0$$

with complex coefficients $a \neq 0$, b, and c has the two solutions

$$z_{1,2} = \frac{-b \pm \sqrt{b^2 - 4ac}}{2a}$$

with no further restriction on the coefficients.

We consider the complex polynomial $f(z) = z^3 - z^2 + z - 1$ and its linear factorization

$$f(z) = (z - 1)(z - i)(z + i).$$

The zeros of f (points z at which f vanishes) are therefore 1 and $\pm i$. (Why are two of them complex conjugates?) The centroid of these three points is given by (15) as $\frac{1}{3}(1 + i - i) = \frac{1}{3}$, i.e., the point $(\frac{1}{3}, 0)$ on the real axis. Consider now the derivative $f'(z) = 3z^2 - 2z + 1$. Its zeros are given by

$$z_{1,2} = \frac{2 \pm \sqrt{4 - 4 \cdot 3}}{2 \cdot 3} = \frac{1 \pm i\sqrt{2}}{2 \cdot 3}.$$

The centroid (midpoint) of these is

$$\frac{1}{2}(z_1 + z_2) = \frac{1}{2}\left[\frac{1 + i\sqrt{2}}{3} + \frac{1 - i\sqrt{2}}{3}\right] = \frac{1}{3},$$

which is the same as the centroid of the zeros of f. Is this a general fact? This question is answered by the following result.

THEOREM 1.8. *Let $f(z) = a_n z^n + \cdots + a_1 z + a_0$ be a complex polynomial of degree n ($a_n \neq 0$). Then the centroid of the zeros of f is the centroid of the zeros of f'.*

Proof: It may be helpful to work out first the case of a degree 2 polynomial $f(z) = a_2 z^2 + a_1 z + a_0$, whose zeros are given by

$$z_{1,2} = \frac{-a_1 \pm \sqrt{a_1^2 - 4a_2 a_0}}{2a_2}$$

with midpoint $\frac{1}{2}(z_1 + z_2) = -\frac{a_1}{2a_2}$. This is indeed the unique zero of the derivative $f'(z) = 2a_2 z + a_1$.

In the general case we use the fact that every complex polynomial has a linear factorization

$$f(z) = a_n(z - z_1) \cdots (z - z_n),$$

where $z_1, \ldots, z_n$ are the (not necessarily distinct) zeros of f. This fact is a consequence of what is called the fundamental theorem of algebra. If we expand this expression by powers of z, the coefficient of z^{n-1} is $-a_n(z_1 + \cdots + z_n)$. But this is the coefficient a_{n-1}. Thus we get the formula

$$z_1 + \cdots + z_n = -\frac{a_{n-1}}{a_n}. \tag{18}$$

Now we do the same for the derivative

$$f'(z) = na_n z^{n-1} + (n-1)a_{n-1}z^{n-2} + \cdots + a_1$$

and its zeros $z'_1, \ldots, z'_{n-1}$. We find for the coefficient of z^{n-2} the expressions $-na_n(z'_1 + \cdots + z'_{n-1})$ and $(n-1)a_{n-1}$, and therefore the formula

$$z'_1 + \cdots + z'_{n-1}, = -\frac{n-1}{n} \cdot \frac{a_{n-1}}{a_n}. \tag{19}$$

Comparing (18) and (19) yields the formula

$$\frac{1}{n}(z_1 + \cdots + z_n) = \frac{1}{n-1}(z'_1 + \cdots + z'_{n-1}),$$

which proves Theorem 1.8.

A repeated application of this argument shows the following fact.

COROLLARY 1.9. *The centroid of the zeros of a polynomial f of degree n and all of its derivatives up to order $n-1$ are the same.*

The $(n-1)^{\text{st}}$ derivative of f is of degree one, and the centroid in question is its unique zero. The next derivative is constant, and the argument does not apply any more.

1.10 Centroid of mass-points. Consider first the following problem. Let $A_1, \ldots, A_p$ be points with centroid U, while V is the centroid of $B_1, \ldots, B_q$ and W is the centroid of $C_1, \ldots, C_r$. What is the relationship of the centroid G of all the points A_i, B_j, C_k to U, V, and W? We calculate

$$G = \frac{1}{p+q+r}\left(\sum_{i=1}^{p} A_i + \sum_{j=1}^{q} B_j + \sum_{k=1}^{r} C_k\right) = \frac{1}{p+q+r}(pU + qV + rW).$$

Thus it seems reasonable to weight the points U, V, W with weights or masses p, q, r. This leads to the following generalization of definition (16).

Let $A_1, \ldots, A_n$ be n points and let $a_1, \ldots, a_n$ be arbitrary numbers (think of masses or weights attached to $A_1, \ldots, A_n$). The pairs (a_i, A_i) are called **weighted points** or **mass-points**. The **centroid** is defined by

$$G = \frac{1}{a_1 + \cdots + a_n}(a_1 A_1 + \cdots + a_n A_n). \tag{20}$$

The situation considered earlier corresponds to unit masses attached to each point. Here, if the a_i's are interpreted as masses, the centroid G has total mass $a_1 + \cdots + a_n$. Equation (20) can be written as

$$(a_1 + \cdots + a_n)G = a_1 A_1 + \cdots + a_n A_n$$

or equivalently as a summation formula for mass-points

$$((a_1 + \cdots + a_n), G) = (a_1, A_1) + \cdots + (a_n, A_n).$$

Note that formula (10) for a point P on the line ℓ_{AB} can be viewed as follows: P is the centroid of the mass-points (a, A) and (b, B) with total mass $a + b = 1$. More generally we can write

$$P = \frac{1}{a+b}(aA + bB) \qquad \text{with } a + b \neq 0. \tag{21}$$

Then P is the centroid of mass $a + b$ of the mass-points (a, A) and (b, B). Formula (21) is equivalent to

$$P - A = \frac{b}{a+b}(B - A) = -\frac{b}{a+b}(A - B)$$

or by symmetry also

$$P - B = \frac{a}{a+b}(A - B).$$

This shows that

$$P - A = -\frac{b}{a}(P - B) \qquad \text{(provided } a \neq 0\text{)}$$

$$\text{or} \tag{22}$$

$$P - B = -\frac{a}{b}(P - A) \qquad \text{(provided } b \neq 0\text{)}.$$

It is convenient to use for these situations the symbolic notation

$$-\frac{b}{a} = \frac{P - A}{P - B}$$

$$-\frac{a}{b} = \frac{P - B}{P - A}. \tag{23}$$

The symbol $\dfrac{P - A}{P - B}$ is not to be interpreted as division of vectors but, by definition, $\dfrac{P - A}{P - B} = -\dfrac{b}{a}$ is an alternate way to write $P - A = -\dfrac{b}{a}(P - B)$. No other meaning is intended.

For the case of three mass-points (a, A), (b, B), (c, C) with $a + b + c \neq 0$ and $a + b \neq 0$, $b + c \neq 0$, $a + c \neq 0$ we consider

$$A' = \frac{bB + cC}{b+c}, \quad B' = \frac{aA + cC}{a+c}, \quad C' = \frac{aA + bB}{a+b}.$$

For $a = b = c = 1$ we obtain the triangle $\triangle A'B'C'$ of midpoints of the triangle $\triangle ABC$. This suggests the following statement generalizing Theorem 1.4.

THEOREM 1.10. *The lines $\ell_{AA'}$, $\ell_{BB'}$ and $\ell_{CC'}$ are concurrent in the centroid G of the mass-points $(a, A), (b, B), (c, C)$.*

Proof: It is sufficient to show that $G \in \ell_{AA'}$ (why?). But

$$G = \frac{1}{a+b+c}(aA + bB + cC) = \frac{a}{a+b+c}A + \frac{b+c}{a+b+c}A'.$$

Since the sum of the coefficients equals one, this proves $G \in \ell_{AA'}$.

Note that the centroid G calculated at the beginning of this section is the centroid of the mass-points $(p, U), (q, V)$, and (r, W).

EXERCISE 1.16. (a) Let $ABCD$ be a quadrilateral. Consider the points

$$E = \frac{1}{3}(A + 2B), \quad F = \frac{1}{3}(2B + C), \quad G = \frac{1}{3}(C + 2D), \quad H = \frac{1}{3}(2D + A)$$

and prove that $EFGH$ is a parallelogram.

(b) Let $(a, A), (b, B), (c, C), (d, D)$ be mass-points. Consider

$$E = \frac{1}{a+b}(aA + bB), \qquad F = \frac{1}{b+c}(bB + cC),$$

$$G = \frac{1}{c+d}(cC + dD), \qquad H = \frac{1}{d+a}(dD + aA).$$

Under which conditions on a, b, c, d are the points E, F, G, H a parallelogram?

1.11 Barycentric coordinates. Let A, B and C be points which do not lie on a line.

THEOREM 1.11. *For every point P of the plane there is a unique representation*

$$P = aA + bB + cC \qquad \text{with } a + b + c = 1. \tag{24}$$

Proof: To clarify the formulas used to prove the existence of such a representation, consider Figure 1.11

Consider the points X and Y, which are the projections of P onto the lines ℓ_{CA} and ℓ_{CB} parallel to the lines ℓ_{CB} and ℓ_{CA}, respectively. In other words, X is the intersection point of the line parallel to ℓ_{CB} through P with the line ℓ_{CA}, and similarly for Y. Then by property (D), expressing the 2-dimensionality of the plane, there are unique numbers x, y such that

$$X - C = x(A - C), \quad Y - C = y(B - C).$$

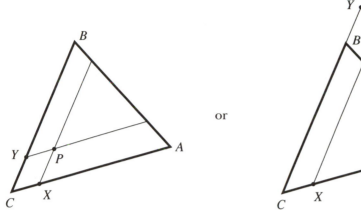

or

Figure 1.11.

Therefore

$$P - C = (X - C) + (Y - C) = x(A - C) + y(B - C) = xA + yB - (x + y)C$$

or with $a = x$, $b = y$, $c = 1 - (x + y)$

$$P = aA + bB + cC \qquad \text{with } a + b + c = 1.$$

To prove the uniqueness of a, b, c in (24), we assume

$$aA + bB + cC = a'A + b'B + c'C$$

and $a + b + c = a' + b' + c' = 1$. Then

$$(a - a')A + (b - b')B + (c - c')C = O$$

and we have to show that $r = a - a' = 0$, $s = b - b' = 0$, $t = c - c' = 0$. Assume e.g., $t \neq 0$, so that

$$C = -\frac{rA}{t} - \frac{sB}{t}.$$

But $-\dfrac{r}{t} - \dfrac{s}{t} = -\dfrac{r + s}{t} = 1$, since $r + s + t = 0$ by the assumptions made. This proves $C \in \ell_{AB}$, contradicting the non-collinearity assumption on A, B, and C. Similarly $s \neq 0$ or $r \neq 0$ leads to a contradiction.

The numbers a, b, c are called the **barycentric coordinates** of P. More generally, for any point P we can write

$$P = \frac{aA + bB + cC}{a + b + c} \tag{25}$$

with any three numbers satisfying $a + b + c \neq 0$. Thus P can be viewed as the centroid of the mass-points (a, A), (b, B), (c, C). If one of the masses is zero, then P is on the line of the opposite edge of the $\triangle ABC$, e.g., $a = 0$ implies $P \in \ell_{BC}$. If all masses are positive, the point P is inside $\triangle ABC$. If all masses are non-negative, the point P is inside or on the edges of $\triangle ABC$.

EXERCISE 1.17. Prove that the points with barycentric coordinates $a = $ constant are the points of a line ℓ parallel to ℓ_{BC}.

EXERCISE 1.18. Paint mixing. Let A be the color red, B the color blue, C the color yellow. If one mixes a quarts of red paint with b quarts of blue paint and c quarts of yellow paint, one gets $a + b + c$ quarts of paint of color defined by the formula

$$\frac{1}{a+b+c}(aA + bB + cC).$$

E.g., for $a = \frac{1}{2}$, $b = \frac{1}{2}$, $c = 0$ the result is one quart of purple paint, for $a = 0$, $b = \frac{1}{2}$, $c = \frac{1}{2}$ the result is one quart of green paint.

Mix one quart of paint of color $\frac{1}{6}(A + 2B + 3C)$ with an unknown quantity of mixed paint using only two colors, so as to obtain an unknown quantity of mixed paint of color $\frac{1}{5}(2A + 2B + C)$. Determine the unknown quantities and color.

As a further application, we consider a complex polynomial of degree 3 with zeros z_1, z_2, z_3.

THEOREM 1.12. *The zeros of the derivative f' have a representation $\sum_{i=1}^{3} a_i z_i$ with $\sum_{i=1}^{3} a_i = 1$, a_i real and $a_i \geq 0$.*

If the zeros z_1, z_2, z_3 are distinct and not collinear, this is an instance of a barycentric representation. In this case the numbers a_i must be unique, and the zeros of f' are points of $\triangle z_1 z_2 z_3$.

Proof: Let ζ be a zero of f'. If $\zeta = z_i$ for some $i = 1, 2, 3$, there is nothing to prove. Thus we can assume that ζ is a zero of f' distinct from the z_i's. Now

$$f(z) = c(z - z_1)(z - z_2)(z - z_3),$$

and thus

$$f'(z) = c[(z - z_2)(z - z_3) + (z - z_1)(z - z_3) + (z - z_1)(z - z_2)].$$

Since $f(\zeta) \neq 0$, this yields

$$\frac{f'(\zeta)}{f(\zeta)} = \frac{1}{\zeta - z_1} + \frac{1}{\zeta - z_2} + \frac{1}{\zeta - z_3} = 0.$$

Multiplying the ith term by $\dfrac{\overline{\zeta - z_i}}{\overline{\zeta - z_i}}$, we find

$$\frac{\overline{\zeta - z_1}}{|\zeta - z_1|^2} + \frac{\overline{\zeta - z_2}}{|\zeta - z_2|^2} + \frac{\overline{\zeta - z_3}}{|\zeta - z_3|^2} = 0,$$

or, after conjugation,

$$\frac{\zeta - z_1}{|\zeta - z_1|^2} + \frac{\zeta - z_2}{|\zeta - z_2|^2} + \frac{\zeta - z_3}{|\zeta - z_3|^2} = 0.$$

This can be written in the form

$$\zeta = \frac{1}{\dfrac{1}{|\zeta - z_1|^2} + \dfrac{1}{|\zeta - z_2|^2} + \dfrac{1}{|\zeta - z_3|^2}} \cdot \left(\frac{z_1}{|\zeta - z_1|^2} + \frac{z_2}{|\zeta - z_2|^2} + \frac{z_3}{|\zeta - z_3|^2} \right),$$

which is of the desired form with

$$a_i = \frac{1}{\dfrac{1}{|\zeta - z_1|^2} + \dfrac{1}{|\zeta - z_2|^2} + \dfrac{1}{|\zeta - z_3|^2}} \cdot \frac{1}{|\zeta - z_i|^2}.$$

By construction, a_i is real and > 0. Further, $\sum a_i = 1$. Note that the case $\zeta = z_i$ corresponds to $a_i = 1$ and $a_j = 0$ for $j \neq i$.

The geometric form of the result above is that every zero ζ of f' is inside $\triangle z_1 z_2 z_3$, provided z_1, z_2, z_3 are non-collinear zeros of f and $\zeta \neq z_i$ ($i = 1, 2, 3$). If ζ_1 and ζ_2 denote the zeros of f', then an interesting question concerns the precise location of ζ_1 and ζ_2. It can be shown that ζ_1 and ζ_2 are the foci of an ellipse inscribed in $\triangle z_1 z_2 z_3$ and having contact with the edges of $\triangle z_1 z_2 z_3$ in the midpoints [M. Marden, Geometry of Polynomials, Math. Surveys no. 3, American Math. Soc. 1966]. An extreme example is the polynomial

$$f(z) = (z - 1) \left(z - e^{\frac{2\pi}{3} i} \right) \left(z - e^{-\frac{2\pi}{3} i} \right).$$

whose zeros 1, $e^{\pm \frac{2\pi}{3} i}$ are located at the vertices of an equilateral triangle. The derivative f' has 0 as a double zero, and the ellipse is in this case the inscribed circle of the triangle. A similar fact holds for the two zeros ζ_1, ζ_2 of the derivative of

$$f(z) = (z - z_1)^{m_1}(z - z_2)^{m_2}(z - z_3)^{m_3}$$

distinct from z_1, z_2, z_3. The points of contact with the edges of $\Delta z_1 z_2 z_3$ of the inscribed ellipse with foci ζ_1, ζ_2 are then the points

$$g_1 = \frac{m_2 z_2 + m_3 z_3}{m_2 + m_3}, \quad g_2 = \frac{m_1 z_1 + m_3 z_3}{m_1 + m_3}, \quad g_3 = \frac{m_1 z_1 + m_2 z_2}{m_1 + m_2}.$$

EXERCISE 1.19. Let $z_1, \ldots, z_n$ be the zeros of a polynomial f of degree n. Prove the following generalizations of Theorem 1.12, using the same proof method. Every zero of the derivative f' has a representation $\sum_{i=1}^{n} a_i z_i$ with $\sum_{i=1}^{n} a_i = 1$, a_i real and ≥ 0. This is a theorem due to Gauss (1835) and Lucas (1874).

1.12 Theorems of Ceva and Menelaus. Let A', B', C' be points on the sides of $\triangle ABC$ (A' on the side opposite A, distinct from the vertices B and C, and similarly for the points B' and C'). What is the condition for the concurrence of the lines $\ell_{AA'}$, $\ell_{BB'}$, $\ell_{CC'}$? The answer is given by the Theorem of Ceva stated below. It is conveniently expressed in terms of the symbolic ratio notation used in (23) for the ratios of the numbers determining the point A' on ℓ_{BC}, namely $\frac{A' - B}{A' - C}$, and similarly for $B' \in \ell_{AC}$ and $C' \in \ell_{AB}$.

THEOREM 1.13   (Theorem of Ceva; Giovanni Ceva, 1648–1734). *The lines* $\ell_{AA'}$, $\ell_{BB'}$, $\ell_{CC'}$ *are concurrent if and only if*

$$\frac{A' - B}{A' - C} \cdot \frac{B' - C}{B' - A} \cdot \frac{C' - A}{C' - B} = -1. \tag{26}$$

Note the cyclical nature of this product: the second ratio is obtained from the first by replacing A', B, C with B', C, A, and similarly for the third.

Proof: We consider first the case where the lines $\ell_{AA'}$, $\ell_{BB'}$, $\ell_{CC'}$ are concurrent in a point G. Let a, b, c be the barycentric coordinates of G, i.e.,

$$G = aA + bB + cC \qquad \text{with } a + b + c = 1.$$

Next we observe that $b + c \neq 0$, for $b + c = 0$ would imply $a = 1$ and

$$G = A + b(B - C) \quad \text{or} \quad G - A = b(B - C).$$

This would imply $\ell_{AG} \,\|\, \ell_{BC}$ (line ℓ_{AG} parallel to line ℓ_{BC}), contradicting $G \in \ell_{AA'}$. It follows that we can write

$$G = aA + (b+c)\frac{bB + cC}{b+c},$$

or

$$\frac{G - aA}{b+c} = \frac{bB + cC}{b+c}. \tag{27}$$

Now the LHS (left-hand side) is a point of the line ℓ_{GA}, since $1 - a = b + c$. The RHS (right-hand side) is a point of ℓ_{BC}. The equality shows that (27) represents the intersection point A', i.e.,

$$A' = \frac{bB + cC}{b+c}. \tag{28}$$

According to (23), we find

$$\frac{A' - B}{A' - C} = -\frac{c}{b}.$$

Similarly (for symmetry reasons)

$$B' = \frac{aA + cC}{a+c} \quad \text{and} \quad \frac{B' - C}{B' - A} = -\frac{a}{c},$$

$$C' = \frac{aA + bB}{a+b} \quad \text{and} \quad \frac{C' - A}{C' - B} = -\frac{b}{a}.$$

It follows that

$$\frac{A' - B}{A' - C} \cdot \frac{B' - C}{B' - A} \cdot \frac{C' - A}{C' - B} = \left(-\frac{c}{b}\right) \cdot \left(-\frac{a}{c}\right) \cdot \left(-\frac{b}{a}\right) = -1. \tag{29}$$

Note that the vanishing of one of the numbers a, b, c would imply that G lies on an edge of $\triangle ABC$.

To prove the converse, we have first to describe $A' \in \ell_{BC}$, etc. By assumption, A' is distinct from B or C, and similarly for B' and C'. Let

$$A' = \frac{bB + cC}{b+c} \quad \text{and thus} \quad \frac{A' - B}{A' - C} = -\frac{c}{b}.$$

Since $A' \neq C$, we have $b \neq 0$. Using the same c, we can write for a unique number a that

$$B' = \frac{aA + cC}{a+c} \quad \text{and thus} \quad \frac{B' - C}{B' - A} = -\frac{a}{c}.$$

Since $B' \neq A$, we have $c \neq 0$. The assumption (26) then implies that

$$\left(-\frac{c}{b}\right) \cdot \left(-\frac{a}{c}\right) \cdot \frac{C' - A}{C' - B} = -1$$

and therefore

$$\frac{C' - A}{C' - B} = -\frac{b}{a}.$$

It follows from (23) that

$$C' = \frac{aA + bB}{a + b}.$$

By Theorem 1.10 we can conclude that the lines $\ell_{AA'}$, $\ell_{BB'}$, $\ell_{CC'}$ are concurrent in

$$G = \frac{1}{a + b + c}(aA + bB + cC).$$

This completes the proof of Ceva's Theorem.

A meticulous reader might wonder why we can divide by $a + b + c$ in the last argument. Let us consider the implications of $a + b + c = 0$. By the construction above we have $a + b \neq 0$, $b + c \neq 0$ and $a + c \neq 0$. Then

$$A - A' = A - \frac{bB + cC}{b + c} = \frac{(b + c)A - bB - cC}{b + c}$$

$$B - B' = B - \frac{aA + cC}{a + c} = \frac{(a + c)B - aA - cC}{a + c}$$

Since by our assumption $b + c = -a$ and $a + c = -b$, we then have

$$A - A' = \frac{-aA - bB - cC}{b + c}$$

$$B - B' = \frac{-aA - bB - cC}{a + c}$$

hence

$$A - A' = \frac{a + c}{b + c}(B - B').$$

It follows that the lines $\ell_{AA'}$, $\ell_{BB'}$ (and by symmetry also $\ell_{CC'}$) are parallel lines. Intuitively, they intersect in a point at infinity.

EXERCISE 1.20. Let $\ell_{AA'}$, $\ell_{BB'}$, $\ell_{CC'}$ be lines concurring in G as in Theorem 1.13. Prove the following identity

$$\frac{G - A}{A' - A} + \frac{G - B}{B' - B} + \frac{G - C}{C' - C} = 2.$$

EXERCISE 1.21. Let $\ell_{AA'}$, $\ell_{BB'}$, $\ell_{CC'}$, be lines concurring in G as in Theorem 1.13. For which G inside $\triangle ABC$ is the number

$$\frac{G-A}{G-A'} + \frac{G-B}{G-B'} + \frac{G-C}{G-C'}$$

a maximum? (Note that each of these numbers is negative.)

To give another interpretation of barycentric coordinates, we consider again the configuration of the three lines $\ell_{AA'}$, $\ell_{BB'}$, $\ell_{CC'}$ concurring in G. As above,

$$G = aA + bB + cC \quad \text{with} \quad a + b + c = 1.$$

THEOREM 1.14. *The areas of the triangles $\triangle GBC$, $\triangle GCA$, $\triangle GAB$ are proportional to the numbers a, b, c (see Figure 1.12).*

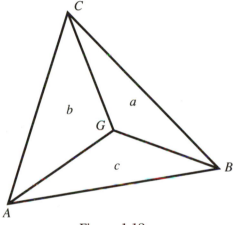

Figure 1.12.

Proof: Let C' be the intersection point of ℓ_{CG} with ℓ_{AB}. First we observe that the areas of $\triangle AC'C$ and $\triangle C'BC$ are proportional to the numbers b and a, since the triangles have the same height, and the length of the bases are proportional to b and a, in this order. It follows that

$$\frac{b}{a} = \frac{\text{area } \triangle CAC'}{\text{area } \triangle CC'B} = \frac{\text{area } \triangle GAC'}{\text{area } \triangle GC'B}$$

$$= \frac{\text{area } \triangle CAC' - \text{area } \triangle GAC'}{\text{area } \triangle CC'B - \text{area } \triangle GC'B}$$

$$= \frac{\text{area } \triangle CAG}{\text{area } \triangle CGB}$$

and similarly for the other ratios.

For the reader who observes that metric notions like length and area appear only from Chapter 3 onwards, we note that the discussion above concerns only ratios of areas, and not the actual areas themselves.

Consider now a choice of points A', B', C' on each side of a triangle $\triangle ABC$, different from the vertices. What is the condition for the collinearity of A', B', C'?

THEOREM 1.15   (Theorem of Menelaus; $\sim$ 100 A.D.). *The points A', B', C' on the sides of $\triangle ABC$ are collinear if and only if*

$$\frac{A' - B}{A' - C} \cdot \frac{B' - C}{B' - A} \cdot \frac{C' - A}{C' - B} = 1. \tag{30}$$

Observe again the cyclic nature of this product. The situation is illustrated by Figure 1.13.

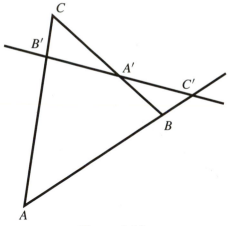

Figure 1.13.

Proof: We consider first the case where the points A', B', C' are collinear. By assumption

$$B' = aA + cC(a + c = 1), \quad A' = rB + sC(r + s = 1).$$

This implies

$$sB' - cA' = asA - crB.$$

Now we observe that $s - c \neq 0$, for $s - c = 0$ would imply $a - r = (1-c) - (1-s) = 0$ and hence $as = cr$. Thus

$$B' - A' = a(A - B),$$

which would imply $\ell_{A'B'} \parallel \ell_{AB}$. But C' is the intersection of these two lines, a contradiction. It follows that we have

$$\frac{s\,B' - c\,A'}{s - c} = \frac{as\,A - cr\,B}{s - c}.$$

The LHS is a point of $\ell_{A'B'}$, the RHS a point of ℓ_{AB} (since $\dfrac{1}{s - c}(as - cr) =$

$\dfrac{1}{s - c}[(1 - c)s - c(1 - s)] = 1$). Thus C', the intersection point of $\ell_{A'B'}$ and ℓ_{AB} is precisely this point, i.e.,

$$C' = \frac{as\,A - cr\,B}{s - c}.$$

By (23) this implies

$$\frac{C' - A}{C' - B} = -\frac{-cr}{as} = \frac{cr}{as}.$$

Further,

$$B' = aA + cC \implies \frac{B' - C}{B' - A} = -\frac{a}{c}$$

$$A' = rB + sC \implies \frac{A' - B}{A' - C} = -\frac{s}{r}.$$

It follows that

$$\frac{A' - B}{A' - C} \cdot \frac{B' - C}{B' - A} \cdot \frac{C' - A}{C' - B} = \left(-\frac{s}{r}\right) \cdot \left(-\frac{a}{c}\right) \cdot \frac{cr}{as} = 1.$$

To prove the converse, let

$$A' = \frac{bB + cC}{b + c}, \quad \text{so} \quad \frac{A' - B}{A' - C} = -\frac{c}{b}.$$

Using the same c, we can then write for a unique number a

$$B' = \frac{aA + cC}{a + c}, \quad \text{so} \quad \frac{B' - C}{B' - A} = -\frac{a}{c}.$$

The assumption (30) implies then

$$\left(-\frac{c}{b}\right) \cdot \left(-\frac{a}{c}\right) \cdot \frac{C' - A}{C' - B} = 1$$

and therefore

$$\frac{C' - A}{C' - B} = \frac{b}{a} = -\frac{(-b)}{a}.$$

This implies

$$C' = \frac{aA - bB}{a - b}.$$

We want to show that $C' \in \ell_{A'B'}$, i.e.,

$$C' = \frac{b'B' - a'A'}{b' - a'} = \frac{a'A' - b'B'}{a' - b'} \qquad \text{for appropriate } a' \text{ and } b'.$$

Comparing the two expression for C' leads to the following attempt: $a' = b + c$, $b' = a + c$. Then $b' - a' = a - b$ and

$$\frac{b'B' - a'A'}{b' - a'} = \frac{1}{a - b}\left((a + c) \cdot \frac{aA + cC}{a + c} - (b + c) \cdot \frac{bB + cC}{b + c} \right)$$

$$= \frac{1}{a - b}(aA - bB) = C'.$$

Thus $C' \in \ell_{A'B'}$.

EXERCISE 1.22. Consider the indicated ratios in Figure 1.14. Determine X as a point of ℓ_{CD} and ℓ_{BE} by applying Menelaus' Theorem.

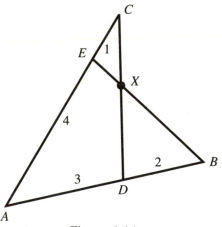

Figure 1.14.

EXERCISE 1.23. Prove Ceva's Theorem as an application of Menelaus' Theorem.

1.13 Theorems of Desargues and Pappus. The following collinearity test for three points A, B, C is useful in many arguments and appears in the proof of the next theorem.

THEOREM 1.16. *Three (distinct) points A, B, C are collinear if and only if there are three numbers a, b, c (not all zero) such that*

$$aA + bB + cC = O \quad \text{and} \quad a + b + c = 0. \tag{31}$$

Proof: If A, B, C are collinear, then $C \in \ell_{AB}$ and $C = aA + bB$ with $a + b = 1$. Setting $c = -(a + b)$, we obtain (31). Assume conversely that (31) holds. One of the a, b, c is not 0. Without loss of generality, we suppose $c \neq 0$. Then

$$C = -\frac{1}{c}(aA + bB).$$

But $-\frac{1}{c}(a + b) = 1$, so $C \in \ell_{AB}$. Similarly if $b \neq 0$ or $a \neq 0$.

To illustrate the statement of the next theorem, consider Figure 1.15.

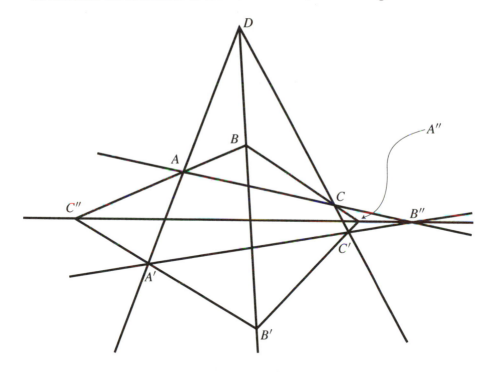

Figure 1.15.

THEOREM 1.17 (Theorem of Desargues; Girard Desargues, 1591–1661**).** *Let $\ell_{AA'}$, $\ell_{BB'}$, $\ell_{CC'}$ be three distinct and concurrent lines with D as the common intersection point. Let*

$$A'' = \ell_{BC} \cap \ell_{B'C'}, \quad B'' = \ell_{AC} \cap \ell_{A'C'}, \quad C'' = \ell_{AB} \cap \ell_{A'B'}$$

be the three intersection points of pairs of corresponding lines as indicated (assuming none of these pairs of lines are parallel). Then A'', B'' and C'' are collinear.

Proof: With the notations of Figure 1.15, we have

$$D = aA + a'A' = bB + b'B' = cC + c'C'$$

with $a + a' = 1$, $b + b' = 1$ and $c + c' = 1$. This implies, e.g.,

$$aA - bB = b'B' - a'A'. \tag{32}$$

Note that $(a - b) + (a' - b') = 0$, so that

$$a - b = b' - a'.$$

We claim that these numbers are different from zero. Assume to the contrary that $a - b = 0$, so $b' - a' = 0$ and $a = b$, $a' = b'$. From (32) we then get

$$a(A - B) = a'(B' - A')$$

and $\ell_{AB} \mathbin{/\mkern-4mu/} \ell_{A'B'}$, contrary to assumptions.

Thus (32) implies

$$\frac{aA - bB}{a - b} = \frac{b'B' - a'A'}{b' - a'}.$$

The LHS is a point of ℓ_{AB}, the RHS a point of $\ell_{A'B'}$, so both sides represent $C'' = \ell_{AB} \cap \ell_{A'B'}$. In particular we obtain the formula

$$aA - bB = (a - b)C''.$$

Similarly we get the formulas

$$bB - cC = (b - c)A''$$

$$cC - aA = (c - a)B''.$$

Adding them all up yields

$$O = (a - b)C'' + (b - c)A'' + (c - a)B''$$

with $0 = (a - b) + (b - c) + (c - a)$ and none of these numbers zero. By Theorem 1.16 the points A'', B'', C'' are collinear.

EXERCISE 1.24. Prove Desargues' Theorem as an application of Menelaus' Theorem. (Hint: The line $\ell_{A'B'}$ is transversal to the sides of $\triangle ABD$, the line $\ell_{A'C'}$ transversal to $\triangle ACD$, the line $\ell_{B'C'}$ transversal to $\triangle BCD$.)

Before proceeding to the next topic, we note the following fact.

PROPOSITION 1.18. *Let A, B be two points with ℓ_{AB} not through O. Then*

$$aA + bB = O \implies a = 0, \; b = 0.$$

Proof: Assume to the contrary that, e.g., $a \neq 0$. Then $A = -\dfrac{b}{a}B$ and $O \in \ell_{AB}$, a contradiction.

This observation appears often in the following form.

COROLLARY 1.19. *Let A, B be two points with ℓ_{AB} not through O. Then*

$$aA + bB = a'A + b'B \implies a = a', \; b = b'.$$

Proof: The assumption implies

$$(a - a')A + (b - b')B = O,$$

to which we can apply Proposition 1.18.

To explain the statement of the next Theorem, we consider Figure 1.16.

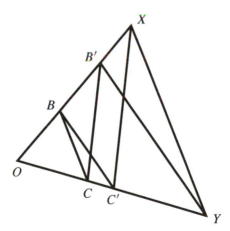

Figure 1.16.

THEOREM 1.20 (Theorem of Pappus; $\sim$ 320 A.D.). *Let $\ell_{BB'}, \ell_{CC'}$ be two lines through O with $X \in \ell_{BB'}, Y \in \ell_{CC'}$. Assume $\ell_{BC'} \; // \; \ell_{B'Y}$ and $\ell_{B'C} \; // \; \ell_{XC'}$. Then $\ell_{BC} \; // \; \ell_{XY}$.*

Proof: The assumption $\ell_{BC'} \mathbin{\!/\mkern-5mu/\!} \ell_{B'Y}$ is expressed by $B' - Y = t(B - C')$. But $B' = rB$ and $Y = sC'$, so $rB - sC' = tB - tC'$. Since $O \notin \ell_{BC'}$, by Corollary 1.19 it follows that $r = s = t$.

Similarly the assumption $\ell_{B'C} \mathbin{\!/\mkern-5mu/\!} \ell_{XC'}$ is expressed by $X - C' = t'(B' - C)$. But $X = r'B'$ and $C' = s'C$, so $r'B' - s'C = t'B' - t'C$, which implies $r' = s' = t'$.

Now we have

$$X = r'B' = r'rB$$
$$Y = sC' = ss'C.$$

Since $rr' = ss' = tt' \ (= a)$, this implies

$$X - Y = a(B - C)$$

and $\ell_{XY} \mathbin{\!/\mkern-5mu/\!} \ell_{BC}$ as claimed.

To summarize what we have done so far: we have discussed a few facts of plane geometry using vector methods. All the topics discussed illustrate the idea of "affine geometry." An expression of this is that in our discussion of triangles the particular shape of the triangle never was significant. This is in contrast to "Euclidean geometry," where length and angle measurements play in addition a significant role. We explain in Chapter 3 how such measurements arise from a scalar product. Before doing this, in Chapter 2 we look at plane geometry from the point of view of groups of transformations.

CHAPTER 2

TRANSLATIONS, DILATATIONS, GROUPS AND SYMMETRIES

In this chapter we introduce a concept allowing us to distinguish between different geometries. This is the concept of a group of transformations. As the simplest examples of transformations we discuss translations and dilatations, and how some geometric figures behave under the effect of these transformations. The composition rules for these transformations leads to the definition of a group. Several examples of groups of symmetries are discussed in complete detail. These are the dihedral groups, the symmetry groups of regular polygons. They will play a prominent role in the discussion of finite groups of isometries in Chapter 4.

2.1 Translations. A *map*, or *correspondence*, α from the set of points of the plane to itself assigns to every point X a well-defined Y, denoted $\alpha(X) = Y$. The map α is *one-to-one* if $\alpha(X) = \alpha(X')$ implies $X = X'$. The map α is *onto*, if for every Y there is (at least) an X such that $\alpha(X) = Y$. A *bijection* (or *permutation* or *transformation*) is a map which is one-to-one and onto.

The first examples are translations. If A is a vector, the *translation by* A is the map from the set of points of the plane to itself given by the formula

$$\tau_A(X) = X + A. \tag{1}$$

If $X = (x_1, x_2)$ and $A = (a_1, a_2)$, then $\tau_A(X) = Y$ is the vector $Y = (y_1, y_2)$ given by $y_1 = x_1 + a_1$, $y_2 = x_2 + a_2$.

PROPOSITION 2.1. *The map τ_A is one-to-one, i.e., $\tau_A(X) = \tau_A(X')$ implies $X = X'$.*

Proof: $X + A = X' + A$ implies $X = X'$.

PROPOSITION 2.2. *The map τ_A is onto, i.e., for every vector Y there is a vector X such that $\tau_A(X) = Y$.*

Proof: The equation $A + X = Y$ has for given Y the (unique) solution $X = Y - A$.

It follows that a translation is a bijection of the plane.

If α and β are maps of the plane to itself, the composition $\beta\alpha$ is the map defined by $(\beta\alpha)(X) = \beta(\alpha(X))$, i.e., first α, then β.

EXERCISE 2.1. Prove that the composition of two bijections is a bijection.

Besides translations, we will deal with other bijections later in this chapter. For the moment we consider the set $\mathcal{T}$ of all translations. The composition $\tau_B\tau_A$ of two translations is calculated by

$$(\tau_B\tau_A)(X) = \tau_B(\tau_A(X)) = \tau_B(A + X) = B + A + X.$$

As a result we have

$$\tau_B\tau_A = \tau_{B+A}. \tag{2}$$

From the commutativity of vector addition we find

$$\tau_B\tau_A = \tau_A\tau_B. \tag{3}$$

This says that the order of composition is irrelevant for translations. This is not necessarily so for more general bijections of the plane. The identity

$$\tau_C(\tau_B\tau_A) = (\tau_C\tau_B)\tau_A \tag{4}$$

is obvious, since both sides are equal to the bijection $\tau_C\tau_B\tau_A$. The composition rule (4) is an associativity law. For the vector O we obtain

$$\tau_O = \iota, \qquad \text{where } \iota = identity \text{ transformation.} \tag{5}$$

The *identity transformation* is formally defined by $\iota(X) = X$ for all X. If α is a bijection, then the inverse α^{-1} is the (unique) bijection such that $\alpha^{-1}\alpha = \iota$, i.e., $\alpha^{-1}(Y)$ is the unique X such that $\alpha(X) = Y$. We have then further $(\alpha\alpha^{-1})(Y) = Y$, so $\alpha\alpha^{-1} = \iota$. For the translation τ_A, the translation τ_{-A} satisfies these properties, hence

$$\tau_A^{-1} = \tau_{-A}. \tag{6}$$

Note that formulas (3), (4), (5), (6) are the exact counterparts of the rules (A1), (A2), (A3), (A4) of vector addition at the beginning of Chapter 1. The correspondence associating to a vector A the translation τ_A changes the rules of vector addition to the composition rules for translations.

A point X is a *fixed point* of a map τ if $\tau(X) = X$.

PROPOSITION 2.3. *For $A \neq O$ the translation τ_A has no fixed point.*

Proof: $\tau_A(X) = X$ implies $A = O$

PROPOSITION 2.4. *A translation τ_A maps a line ℓ into a line $\tau_A(\ell)$ parallel to ℓ.*

Proof: If $\ell = \ell_{PQ}$ and $X \in \ell_{PQ}$, then

$$X = pP + qQ \qquad \text{with } p + q = 1.$$

It follows that

$$\tau_A(X) = pP + qQ + A = p(P + A) + q(Q + A) = p\tau_A(P) + q\tau_A(Q).$$

This shows that $\tau_A(X)$ is a point of the line ℓ through $\tau_A(P)$ and $\tau_A(Q)$. This is the line $\tau_A(\ell)$. It remains to show that $\tau_A(\ell) \mathbin{/\!/} \ell$. This follows from

$$\tau_A(Q) - \tau_A(P) = (Q + A) - (P + A) = Q - P.$$

If the vector A is in the direction of the line ℓ, then $\tau_A(\ell)$ and ℓ coincide.

PROPOSITION 2.5. *Given two points B, C there is a unique translation τ_A such that $\tau_A(B) = C$.*

Proof: The equation to solve is $A + B = C$ for given B and C. The unique solution is $A = C - B$.

As an application of translations, consider the situation of Desargues' Theorem 1.17, but with parallel lines $\ell_{AA'}$, $\ell_{BB'}$ and $\ell_{CC'}$. If we assume $\ell_{AB} \mathbin{/\!/} \ell_{A'B'}$, $\ell_{BC} \mathbin{/\!/} \ell_{B'C'}$, then the conclusion is $\ell_{AC} \mathbin{/\!/} \ell_{A'C'}$. This is easy to verify using Proposition 2.4 applied, e.g., to the translation $\tau_{A'-A}$.

2.2 Central dilatations. If r is a non-zero number, the **central dilatation** with center O and dilatation factor, or ratio, $r \neq 0$ is the map δ_r of the plane to itself given by the formula

$$\delta_r(X) = rX. \tag{7}$$

If $X = (x_1, x_2)$, then $\delta_r(X) = Y$ is the vector $Y = (y_1, y_2)$ given by $y_1 = rx_1$, $y_2 = rx_2$. If $r > 1$, the map δ_r can be thought of as a blow-up. If $0 < r < 1$, the map δ_r is a shrinking to a smaller scale.

PROPOSITION 2.6. *The map δ_r is one-to-one.*

Proof: $\delta_r(X) = \delta_r(X')$ or $rX = rX'$ implies $r(X - X') = O$. Since $r \neq 0$, we have $X - X' = O$ and $X = X'$.

PROPOSITION 2.7. *The map δ_r is a map onto the plane.*

Proof: For given Y, the equation to solve is $rX = Y$. The (unique) solution is $X = 1/(rY)$.

Thus every central dilatation δ_r is a bijection of the plane. The composition $\delta_s \delta_r$ of two central dilatations (first δ_r, then δ_s) is calculated by

$$(\delta_s \delta_r)(X) = s(rX) = srX.$$

It follows that

$$\delta_s \delta_r = \delta_r \delta_s = \delta_{rs}. \tag{8}$$

Note that

$$\delta_1 = \iota \text{ (identity).} \tag{9}$$

From (8), (9) it follows that

$$\delta_r^{-1} = \delta_{1/r}. \tag{10}$$

Next we compose translations and central dilatations. We will see that composition is in general not commutative, i.e., if α is composed with γ, then $\gamma\alpha$ and $\alpha\gamma$ do not have the same meaning.

PROPOSITION 2.8. *For $r \neq 0$ and any vector A we have*

$$\delta_r \tau_A = \tau_{rA} \delta_r. \tag{11}$$

In particular, central dilatations and translations do not commute.

Proof: For any X we have

$$(\delta_r \tau_A)(X) = r(X + A) = rX + rA = \tau_{rA}(rX) = (\tau_{rA} \delta_r)(X).$$

An equivalent formulation of (11) is the formula

$$\delta_r \tau_A \delta_r^{-1} = \tau_{rA}. \tag{12}$$

The expression on the LHS is called the **conjugation** of the translation τ_A by the central dilatation δ_r. We will encounter many conjugation formulas in this book.

The central dilatations considered so far have the origin as the center of dilatation. Suppose now that C is any point. Then the formula

$$\delta_{C,r}(X) = C + r(X - C) = (1 - r)C + rX \tag{13}$$

for $r \neq 0$ defines a map of the plane to itself, which dilatates the plane with center C and dilatation factor r (see Figure 2.1).

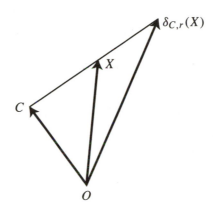

Figure 2.1.

For $C = (c_1, c_2)$, $X = (x_1, x_2)$, and $\delta_{C,r}(X) = Y = (y_1, y_2)$ formula (13) is equivalent to the equations

$$y_1 = c_1 + r(x_1 - c_1) = (1 - r)c_1 + rx_1$$
$$y_2 = c_2 + r(x_2 - c_2) = (1 - r)c_2 + rx_2.$$

Formula (13) can be equivalently expressed by the conjugation formula

$$\delta_{C,r} = \tau_C \delta_r \tau_C^{-1}. \tag{14}$$

To verify this, we calculate

$$(\tau_C \delta_r \tau_C^{-1})(X) = (\tau_C \delta_r)(X - C) = \tau_C(r(X - C)) = C + r(X - C),$$

which coincides with (13). For $C = O$ we have $\delta_{C,r} = \delta_r$.

EXERCISE 2.2. Prove that as a consequence of (14), the map $\delta_{C,r}$ is necessarily a bijection.

Further note that

$$\delta_{C,s}\delta_{C,r} = \delta_{C,sr} \tag{15}$$

$$\delta_{C,1} = \iota \tag{16}$$

$$\delta_{C,r}^{-1} = \delta_{C,1/r}. \tag{17}$$

Formula (14) has the following generalization:

$$\delta_{A+C,r} = \tau_C \delta_{A,r} \tau_C^{-1}. \tag{18}$$

To verify this, we use (14) twice:

$$\tau_C \delta_{A,r} \tau_C^{-1} = \tau_C \tau_A \delta_r \tau_A^{-1} \tau_C^{-1} = \tau_{A+C} \delta_r \tau_{A+C}^{-1} = \delta_{A+C,r}.$$

Next we discuss particular properties of central dilatations.

PROPOSITION 2.9. (i) C *is a fixed point of* $\delta_{C,r}$; (ii) C *is the only fixed point of* $\delta_{C,r}$ *for* $r \neq 1$.

Proof: (i) $\delta_{C,r}(C) = C + r(C - C) = C$. (ii) Let X be a fixed point of $\delta_{C,r}$: $\delta_{C,r}(X) = X$. Then $C + r(X - C) = X$. It follows that $r(X - C) = X - C$ or $(r - 1)(X - C) = O$. For $r \neq 1$ it follows that $X - C = O$.

PROPOSITION 2.10. *A central dilatation* $\delta_{C,r}$ *maps a line* ℓ *into a line* $\delta_{C,r}(\ell)$ *parallel to* ℓ.

Proof: Since the property holds for translations, by (14) it suffices to prove it for central dilatations with center $C = O$. If ℓ is a line through O, then $\delta_r(\ell) = \ell$. Thus we can assume that ℓ is a line not through O. Let $\ell = \ell_{XY}$, so that $P \in \ell_{XY}$ is given by

$$P = xX + yY \qquad \text{with } x + y = 1.$$

Then

$$\delta_r(P) = r(xX + yY) = x\delta_r(X) + y\delta_r(Y).$$

This shows that $\delta_r(P)$ is a point of the line through $\delta_r(X)$, $\delta_r(Y)$. Further

$$\delta_r(Y) - \delta_r(X) = r(Y - X),$$

which proves that $\delta_r(\ell)$ is parallel to ℓ.

The reader will recall that we had verified that the same property holds for a translation. Central dilatation and translations are examples of **collineations**, which are defined as bijections mapping lines into lines.

EXERCISE 2.3. Let A, B be distinct points. Describe the location of all points P such that $P = aA + bB$ with $a + b = r$ (r a constant different from 0).

PROPOSITION 2.11. *Given two points* A, B *on the line* ℓ *through* C *(* $A \neq C$, $B \neq C$ *), there exists a unique central dilatation* $\delta_{C,r}$ *with* $\delta_{C,r}(A) = B$.

Proof: We wish to have $C + r(A - C) = B$, so that necessarily $r(A - C) = B - C$. But there is a unique $r \neq 0$ with this property. For this r the corresponding central dilatation $\delta_{C,r}$ has the desired property.

As shown in the proof, the number r is given by the ratio

$$r = \frac{B - C}{A - C} \tag{19}$$

EXERCISE 2.4. Consider $\triangle ABC$ and the triangle $A'B'C'$ of its midpoints. (i) Consider the dilatation $\delta_{A,2}$ and its effect on $G = \ell_{BB'} \cap \ell_{CC'}$ to conclude that G is also a point of $\ell_{AA'}$ (which proves again the concurrence of the medians of a triangle). (ii) What is the image of $\triangle ABC$ under the central dilatation $\delta_{G,-1/2}$? Describe the result if one repeats this process a large number of times.

EXERCISE 2.5. Prove the following borderline case of Desargues' Theorem 1.17, using a central dilatation. Let $\ell_{AA'}$, $\ell_{BB'}$, $\ell_{CC'}$ be three lines concurrent in D. Assume $\ell_{AB} \mathbin{/\!/} \ell_{A'B'}$ and $\ell_{BC} \mathbin{/\!/} \ell_{B'C'}$. Prove that $\ell_{AC} \mathbin{/\!/} \ell_{A'C'}$. The situation is illustrated in Figure 2.2.

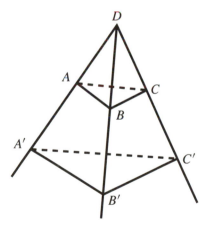

Figure 2.2.

EXERCISE 2.6. Prove Pappus' Theorem 1.20 using central dilatations.

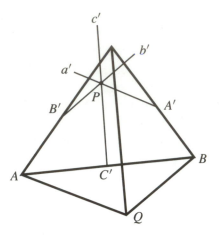

Figure 2.3.

The next property to be discussed is illustrated in Figure 2.3. We consider $\triangle ABC$, its triangle of midpoints $\triangle A'B'C'$, and any point Q. Then we consider line a' parallel to ℓ_{AQ} through A', line b' parallel to ℓ_{BQ} through B', and line c' parallel to ℓ_{CQ} through C'.

THEOREM 2.12   (Theorem of Snapper).

(i) *The lines a', b', c' are concurrent (in a point P).*
(ii) *The centroid G of $\triangle ABC$ lies on ℓ_{PQ} and*

$$\frac{G - P}{G - Q} = -\frac{1}{2}.$$

To explain the interest of this result, we anticipate the concepts of orthocenter (concurrence point of altitudes) and circumcenter (concurrence point of perpendicular bisectors) explained in Chapter 3. If Q is the orthocenter of $\triangle ABC$, then the construction above leads to the circumcenter.

Proof of Theorem 2.12: The central dilatation $\delta_{G,-1/2}$ maps $\triangle ABC$ into $\triangle A'B'C'$ and thus ℓ_{AQ} to a', ℓ_{BQ} to b', and ℓ_{CQ} to c'. The concurrence point Q is therefore mapped to the concurrence point P of a', b', c', which proves (i). By construction, $G \in \ell_{PQ}$ and moreover

$$P - G = -\frac{1}{2}(Q - G),$$

which proves (ii).

EXERCISE 2.7. Figure 2.3 can also be viewed as the construction of Q starting with P. Consider line a parallel to $\ell_{A'P}$ through A, b parallel to $\ell_{B'P}$ through B, and c parallel to $\ell_{C'P}$ through C. Prove that a, b, c are concurrent (in a point Q) with $\dfrac{G-P}{G-Q} = -\dfrac{1}{2}$.

2.3 Central Reflections. The *central reflection* in C is by definition the bijection

$$\sigma_C = \delta_{C,-1}. \tag{20}$$

By (13) this means that

$$\sigma_C(X) = 2C - X. \tag{21}$$

Another way to look at this formula is to write it as

$$C = \frac{1}{2}(X + \sigma_C(X)), \tag{22}$$

i.e., C is the midpoint of X and $\sigma_C(X)$, for each X. This justifies the name central reflection in C. For $C = O$ we have in particular

$$\sigma_O(X) = -X.$$

PROPOSITION 2.13. $\sigma_C^2 = \sigma_C\sigma_C = \iota$.

Proof: For every X we have

$$\sigma_C^2(X) = \sigma_C(\sigma_C(X)) = \sigma_C(2C - X) = 2C - (2C - X) = X.$$

A transformation α which is not the identity, and which satisfies $\alpha^2 = \iota$, is called an *involution*. Note that for an involution α we have $\alpha^{-1} = \alpha$.

What are the fixed points of the involution σ_C? Certainly C is fixed, since

$$\sigma_C(C) = 2C - C = C.$$

PROPOSITION 2.14. C *is the unique fixed point of* σ_C.

This has been proved more generally in Proposition 2.9. We repeat the argument in the special case at hand. If X is a point with $\sigma_C(X) = X$, then $2C - X = X$ and hence $C = X$.

Now we calculate the composition of two central reflections σ_P and σ_Q. For any point X,

$$(\sigma_Q\sigma_P)(X) = \sigma_Q(2P - X) = 2Q - (2P - X) = X + 2(Q - P),$$

which is the same as for the translation $\tau_{2(Q-P)}$.

THEOREM 2.15. *The composition of two central reflections is a translation:*

$$\sigma_Q \sigma_P = \tau_A \qquad \text{with } A = 2(Q - P). \tag{23}$$

Note that

$$\sigma_P \sigma_Q = \tau_B \qquad \text{with } B = 2(P - Q) = -A,$$

and therefore

$$\sigma_P \sigma_Q = \tau_{-A} = \tau_A^{-1} = (\sigma_Q \sigma_P)^{-1}. \tag{24}$$

EXERCISE 2.8. Let α, β be bijections with inverses α^{-1}, β^{-1}. Prove that

$$(\beta\alpha)^{-1} = \alpha^{-1}\beta^{-1}.$$

For the case of involutions α, β this implies $(\beta\alpha)^{-1} = \alpha\beta$.

Is every translation τ_A the composition of two central reflections $\sigma_Q \sigma_P$? The necessary condition is, by (23),

$$A = 2(Q - P). \tag{25}$$

Given any P, we can find a unique Q satisfying this condition by

$$Q = \frac{1}{2}(2P + A) = P + \frac{1}{2}A.$$

Thus we can write τ_A as the composition $\sigma_Q \sigma_P$ in infinitely many ways, but Q is uniquely determined for prescribed P.

EXERCISE 2.9. Write the translation τ_A as the composition $\sigma_Q \sigma_P$ for prescribed Q. Determine the unique P satisfying this condition.

Now we calculate the composition of three central reflections $\sigma_A, \sigma_B, \sigma_C$. For any X,

$$(\sigma_C \sigma_B \sigma_A)(X) = (\sigma_C \sigma_B)(2A - X) = \sigma_C(2B - (2A - X))$$
$$= 2C - (2B - 2A + X) = 2(A + C - B) - X.$$

This is again a formula for a central reflection σ_D with

$$D = A + C - B.$$

Note that this formula can be written $B + D = A + C$, which is precisely the condition that the points A, B, C, D are the four vertices of a parallelogram. We have proved the following fact.

THEOREM 2.16. *The composition $\sigma_C\sigma_B\sigma_A$ of three central reflections is the central reflection σ_D in the fourth vertex of the parallelogram $ABCD$.*

COROLLARY 2.17. *For three points A, B, C we have*

$$\sigma_C\sigma_B\sigma_A = \sigma_A\sigma_B\sigma_C. \tag{26}$$

Proof: Both compositions equal σ_D, where D is the fourth point in the parallelogram $ABCD$.

EXERCISE 2.10. Use central reflections to prove that there is a unique triangle $\triangle ABC$ such that the midpoints of its sides are a given triangle $\triangle A'B'C'$. An alternative argument uses a central dilatation.

EXERCISE 2.11. Verify the conjugation formula

$$\tau_A\sigma_P\tau_A^{-1} = \sigma_Q \qquad \text{for } A = Q - P.$$

Next we calculate the composition of a central reflection σ_P and a translation τ_A. For any X we have

$$(\tau_A\sigma_P)(X) = 2P - X + A = 2(P + \tfrac{1}{2}A) - X$$

which shows that

$$\tau_A\sigma_P = \sigma_Q \qquad \text{with } Q = P + \tfrac{1}{2}A. \tag{27}$$

Similarly

$$(\sigma_P\tau_A)(X) = 2P - (X + A) = 2(P - \tfrac{1}{2}A) - X$$

shows that

$$\sigma_P\tau_A = \sigma_R \qquad \text{with } R = P - \tfrac{1}{2}A. \tag{28}$$

In this particular situation we observe again that $\sigma_P\tau_A$ and $\tau_A\sigma_P$ are distinct unless $A = O$.

2.4 Dilatations. A *dilatation* is defined as a bijection that is either a translation or a central dilatation. A *collineation* is defined as a bijection that maps lines into lines. By Propositions 2.4 and 2.10 a dilatation is a collineation, and maps every line to a parallel line. Is the composition of dilatations necessarily a dilatation? We recall the property (11)

$$\delta_r\tau_A = \tau_{rA}\delta_r, \tag{29}$$

and again point out the importance of paying careful attention to the order of composition of dilatations. For the composition in the reverse order, we ask if there is a C such that $\tau_A \delta_r = \delta_{C,r}$. The answer is that

$$\tau_A \delta_r = \delta_{\frac{1}{1-r}A,r} \qquad \text{for } r \neq 1. \tag{30}$$

Proof of (30)*:* For any X we have

$$(\tau_A \delta_r)(X) = A + rX = \left(\frac{1}{1-r} - \frac{r}{1-r} \right) A + rX$$

$$= \frac{1}{1-r}A + r\left(X - \frac{1}{1-r}A \right),$$

which by (13) is precisely the formula for $\delta_{\frac{1}{1-r}A,r}(X)$.

More generally we claim that

$$\tau_A \delta_{C,r} = \delta_{\frac{1}{1-r}A+C,r} \qquad \text{for } r \neq 1. \tag{31}$$

To prove this, we use consecutively (14), (30), and (18):

$$\tau_A \delta_{C,r} = \tau_A \tau_C \delta_r \tau_C^{-1} = \tau_C \tau_A \delta_r \tau_C^{-1}$$
$$= \tau_C \delta_{\frac{1}{1-r}A,r} \tau_C^{-1} = \delta_{\frac{1}{1-r}A+C,r}.$$

Once we know that $\tau_A \delta_{C,r}$ for $r \neq 1$ is a central dilatation $\delta_{P,r}$, it is easy to find its center P as the (unique) fixed point of $\tau_C \delta_{C,r}$. Namely this condition reads

$$A + C + r(P - C) = P,$$

which implies the formula

$$P = \frac{1}{1-r}A + C$$

as in (31).

We further note that by (14) and (29),

$$\delta_{C,r} \tau_A = \tau_C \delta_r \tau_C^{-1} \tau_A = \tau_C (\tau_r C)^{-1} \delta_r \tau_A$$
$$= (\tau_{(1-r)C}) \tau_{rA} \delta_r = (\tau_{(1-r)C+rA}) \delta_r, \tag{32}$$

which by (30) can again be written as a central dilatation (for $r \neq 1$).

THEOREM 2.18. *The inverse of a dilatation is a dilatation. The composition of two dilatations is a dilatation.*

Proof: The inverse of a translation is a translation, the inverse of a central dilatation a central dilatation, which proves the first statement. The composition of two translations is a translation. The composition of a translation and a central dilatation (different from the identity) is a central dilatation by formulas (29), (30). It remains to prove that the composition of two central dilatations is either a translation or a central dilatation. For this purpose we prove first the formula

$$\delta_{B,s}\delta_{A,r} = \tau_{(1-s)B+s(1-r)A}\delta_{sr}. \tag{33}$$

By (14) and (29), applied twice we have

$$
\begin{aligned}
\delta_{B,s}\delta_{A,r} &= (\tau_B \delta_s \tau_B^{-1})(\tau_A \delta_r \tau_A^{-1}) \\
&= (\tau_B \tau_{sB}^{-1} \delta_s)(\tau_A \tau_{rA}^{-1} \delta_r) \\
&= \tau_{(1-s)B} \tau_{s(1-r)A} \delta_s \delta_r \\
&= \tau_{(1-s)B+s(1-r)A} \delta_{sr}.
\end{aligned}
$$

If $sr = 1$, let $C = (1-s)B + (s-sr)A = (s-1)(A-B)$. Then

$$\delta_{B,s}\delta_{A,r} = \tau_C, \qquad \text{i.e., a translation.}$$

If $sr \neq 1$, then by (30)

$$\delta_{B,s}\delta_{A,r} = \tau_C \delta_{sr} = \delta_{\frac{1}{1-sr}C,sr}. \tag{34}$$

This completes the proof of Theorem 2.18.

From (34) we can conclude that

$$P = \frac{1}{1-sr}C = \frac{1}{1-sr}[s(1-r)A + (1-s)B]$$

is the center of the resulting central dilatation.

Again we observe that the center P can be obtained easily as the unique fixed point of $\delta_{B,s}\delta_{A,r}$, once we know that this transformation is a central dilatation $\delta_{P,sr}$. The sum of the coefficients of A and B equals

$$\frac{1}{1-sr}[s(1-r) + (1-s)] = 1$$

and therefore (still assuming $sr \neq 1$),

$$\delta_{B,s}\delta_{A,r} = \delta_{P,sr} \qquad \text{with } P \in \ell_{AB}. \tag{35}$$

It is of interest to calculate the composition of two dilatations of the form $\delta = \tau_A \delta_r$ and $\delta' = \tau_{A'} \delta_{r'}$. Using (29) we find

$$\delta\delta' = \tau_A \delta_r \tau_{A'} \delta_{r'} = \tau_A \tau_{rA'} \delta_r \delta_{r'} = \tau_{A+rA'} \delta_{rr'}.$$

If we represent δ by the pair (A, r) and δ' by (A', r'), we can write this formula as

$$(A, r) \cdot (A', r') = (A + rA', rr'), \tag{36}$$

where the LHS is a symbolic product (often called semi-direct product) corresponding to the composition $\delta\delta'$. Note that $\delta\delta' = \delta'\delta$ exactly if $A + rA' = A' + r'A$ or equivalently

$$(1 - r')A = (1 - r)A',$$

which in general is not the case.

EXERCISE 2.12. Prove the formula

$$\delta_{C,r} \tau_A = \tau_{rA} \delta_{C,r}.$$

EXERCISE 2.13. Let $r \neq 1$ and $s \neq 1$. Use formula (33) to show

$$\delta_{B,s} \delta_{A,r} = \delta_{A,r} \delta_{B,s} \quad \text{implies} \quad A = B.$$

EXERCISE 2.14. Let ℓ be a transversal to the triangle $\triangle ABC$ as in Figure 1.13 with intersection points A', B', C'. Prove the Menelaus condition

$$\frac{A' - B}{A' - C} \cdot \frac{B' - C}{B' - A} \cdot \frac{C' - A}{C' - B} = 1$$

using dilatations. (Hint: Let $\dfrac{1}{r} = \dfrac{A' - B}{A' - C}$ and calculate $\delta_{A',r}(B)$. The argument is the same for the other ratios, $\dfrac{1}{s}$ and $\dfrac{1}{t}$. Then examine the composition of the three resulting central dilatations. What are its fixed points?)

2.5 Groups of transformations. We have encountered several sets of bijections or transformations of the plane with the property that the composition of two transformations is again a transformation of the set, and the inverse of a transformation is a transformation of the set.

DEFINITION. A set $\mathcal{G}$ of transformations of a (non-empty) set $\mathcal{V}$ is a *group of transformations* if the following properties hold:

(i) if α, β are in $\mathcal{G}$, then the composition $\beta\alpha$ is in $\mathcal{G}$,
(ii) if α is in $\mathcal{G}$, the inverse α^{-1} is in $\mathcal{G}$.

Note that α and α^{-1} in $\mathcal{G}$ implies that the identity transformation $\iota = \alpha^{-1}\alpha$ is also in $\mathcal{G}$. The composition of transformations is naturally associative, i.e., $\gamma(\beta\alpha) = (\gamma\beta)\alpha$, for both of these transformations have exactly the same effect.

If $\mathcal{V}$ denotes the points of the plane, then the set $\mathcal{T}$ of translations is a group. So is the set of central dilatations with the same center C. By Theorem 2.18 the set $\mathcal{D}$ of all dilatations is a group. Further, the set of all transformations of $\mathcal{V}$ is a group. On the other hand the set of all central reflections is not a group (see Theorem 2.15).

If $\mathcal{V}$ is any set, the group of all transformations or bijections of $\mathcal{V}$ is also called the *group of permutations* of $\mathcal{V}$.

It is important to note that the composition $\beta\alpha$ of two transformations need not equal the composition $\alpha\beta$. A group $\mathcal{G}$ of transformations is *commutative* or *abelian* (named after the Norwegian mathematician N. H. Abel), if $\beta\alpha = \alpha\beta$ for all $\alpha, \beta \in \mathcal{G}$. An example is the group of translations, or the group of central dilatations with the same center. The group $\mathcal{D}$ of all dilatations is not commutative, as we have seen earlier.

A property is said to be *invariant* under a group $\mathcal{G}$ of transformations if the property still holds after the transformations of $\mathcal{G}$ are applied. A geometric figure is *invariant* under $\mathcal{G}$ if it is mapped to itself by the transformations of $\mathcal{G}$. A related concept is the concept of equivalence under the transformations of $\mathcal{G}$ to be defined as follows.

DEFINITION. Two figures F_1, F_2 in the plane are *related by a group $\mathcal{G}$ of transformations* if there exists a transformation α in $\mathcal{G}$ mapping F_1 into F_2, i.e., $\alpha(F_1) = F_2$. In symbols we write

$$F_1 \sim F_2 \quad \text{with respect to} \quad \mathcal{G}$$

or simply $F_1 \sim F_2$ ($\mathcal{G}$ being tacitly understood).

This relation has the following properties.

Transitivity: If $F_1 \sim F_2$ and $F_2 \sim F_3$, then $F_1 \sim F_3$.

The proof consists in the observation that if $\alpha(F_1) = F_2$ and $\beta(F_2) = F_3$ for some α, β in $\mathcal{G}$, then $(\beta\alpha)(F_1) = F_3$ with $\beta\alpha$ in $\mathcal{G}$.

Symmetry: If $F_1 \sim F_2$, then $F_2 \sim F_1$.

The proof consists in the statement that if $\alpha(F_1) = F_2$, then $\alpha^{-1}(F_2) = F_1$ with α^{-1} in $\mathcal{G}$. To this property is added the trivial sounding property

Reflexivity: $F_1 \sim F_1$.

This is equivalent to stating that the identity transformation ι is in $\mathcal{G}$.

A relation with the property of transitivity, symmetry and reflexivity is called an **equivalence relation**. The important fact to recognize is that these properties correspond precisely to the properties defining a group of transformations. Instead of saying that two figures are related by a group $\mathcal{G}$, we say then that they are equivalent under $\mathcal{G}$.

Consider, e.g., the group $\mathcal{D}_C$ of central dilatations with a fixed center C. Two figures F_1, F_2 are equivalent under $\mathcal{D}_C$ if $\delta_{C,r}(F_1) = F_2$ for some $r \neq 0$. Then also $F_1 = \delta_{C,1/r}(F_2)$ (symmetry of the relation). To mention the reflexivity property in concrete contexts seems overly pedantic.

Equivalence under the group of translations $\mathcal{T}$ is so obvious as to be almost embarrassing to formulate in the group context. Do it anyway as an exercise in the uses of this terminology.

Klein formulated in his Erlanger Program of 1872 the following idea: to each group of transformations of the plane corresponds a geometry. Groups of transformations had been used in geometry before. The originality of this idea of Felix Klein and Sophus Lie was to reverse the order and to make the group the primary object of attention.

In this sense there is a dilatational geometry associated to the group of dilatations $\mathcal{D}$. There is a translational geometry associated to the group of translations $\mathcal{T}$ (and usually considered too trivial to be worth mentioning). Affine geometry in the spirit of Chapter 1 can now be defined as the geometry associated to the group of collineations (recall that these are the bijections mapping lines to lines). The topic of Chapter 4 will be to discuss Euclidean geometry as the geometry associated to the group of isometries of the plane. This is a point of view which has proved very useful not only in geometry, but also in other branches of mathematics and in physics. The group of transformations mapping solutions of an equation to solutions is an essential object associated to the equation. E.g., in atomic physics, the regularities in the periodic table of elements are a direct consequence of an atomic model invariant under (spatial) rotations.

The idea of symmetry is very much related to the idea of equivalence under the transformations of a group. Usually the symmetries of a figure are the transformations leaving this figure invariant. They form a group of transformations. If we identify figures equivalent under a group $\mathcal{G}$, as we subconsciously do, then the transformations in $\mathcal{G}$ become symmetries of the (equivalence class of) figures. In this sense the ideas of symmetry and groups of transformations are identical. Note that the properties of transitivity, symmetry and reflexivity are automatically associated with the concept of symmetry. A delightful book explaining the ideas of symmetry and transformation groups to a wider audience is "Symmetry", by Hermann Weyl, Princeton University Press (1952).

2.6 Abstract groups. It has proved fruitful to abstract the properties characterizing a group of transformations and use them to define a group in a way which does not depend on transformations or geometry and in which the elements of the group used need not be transformations. For two transformations, composition is well-defined. For two numbers, composition might be addition or multiplication. In general we simply postulate a composition operation. This leads to the following concept of an (abstract) group.

DEFINITION. A *group* $\mathcal{G}$ is a (non-empty) set, in which there is defined a rule of composition denoted XY or $X \cdot Y$ for elements X and Y of $\mathcal{G}$ (XY is a further element of $\mathcal{G}$) and satisfying the following properties.

(G1) There exists an element E in $\mathcal{G}$, called the identity, such that $EX = XE = X$ for all X in $\mathcal{G}$.

(G2) For every X in $\mathcal{G}$, there is an element X^{-1} in $\mathcal{G}$, called the inverse of X, such that $X^{-1}X = XX^{-1} = E$.

(G3) For all X, Y and Z in $\mathcal{G}$ we have $X(YZ) = (XY)Z$ (associativity).

If $\mathcal{G}$ is group of transformations, it is a group with the composition rule being the composition of transformations.

A group $\mathcal{G}$ is *commutative* or *abelian* if $XY = YX$ for all X, Y in $\mathcal{G}$.

Here are a few examples of commutative groups. The real numbers $(\mathbf{R}, +)$ with respect to addition as composition, and with 0 as the identity element. The nonzero real numbers $(\mathbf{R}^*, \cdot)$ with respect to multiplication as composition, with 1 as the identity element. The set $\mathcal{V}$ of all vectors of the plane with addition is a commutative group. The set $\mathcal{T}$ of all translations of the plane is a group which in substance (abstractly) looks very much like $\mathcal{V}$.

EXERCISE 2.15. Let $\mathcal{G}$ be a group. Prove that if there are two elements E, E' both satisfying (G1), then $E = E'$. This justifies speaking of *the* identity of a group $\mathcal{G}$. Note that for a group of transformations this nitpicking is not necessary.

EXERCISE 2.16. Let $\mathcal{G}$ be a group. Assume that for some X in $\mathcal{G}$ there are two elements Y, Y' satisfying $YX = XY = E$ and $Y'X = XY' = E$. Prove that $Y = Y'$. This justifies speaking of *the* inverse X^{-1} of an element X of $\mathcal{G}$.

EXERCISE 2.17. Let $\mathcal{G}$ be a group. Prove that for any X we have $(X^{-1})^{-1} = X$. Prove that for any X and Y we have $(XY)^{-1} = Y^{-1}X^{-1}$.

EXERCISE 2.18. Let $\mathbf{Z}_2$ be the set consisting of two elements 0, 1. Verify that the composition rule $0 + 0 = 0$, $0 + 1 = 1$, $1 + 0 = 1$, $1 + 1 = 0$ turns $\mathbf{Z}_2$ into a group with 0 as identity element. Which element is the inverse of 1?

The example in Exercise 2.18 can be generalized as follows. Let $\mathbf{Z}_n = \{0, 1, \ldots, n - 1\}$. The composition rule is given in form of the table in Figure 2.4.

+	0	1	2	$\ldots$	$n - 2$	$n - 1$
0	0	1	2	$\ldots$	$n - 2$	$n - 1$
1	1	2	3	$\ldots$	$n - 1$	0
2	2	3	4	$\ldots$	0	1
$\vdots$	$\vdots$	$\vdots$	$\vdots$		$\vdots$	$\vdots$
$n - 1$	$n - 1$	0	1	$\ldots$	$n - 3$	$n - 2$

Figure 2.4.

This group is called the group of residues left after division by n, or the **group of integers mod** n (read this as "modulo n"). It is a commutative group with 0 as the identity. Note that each row and column contains each element of $\mathbf{Z}_n$ exactly once. The inverse of 2, e.g., is $n - 2$, since $2 + (n - 2) = 0$ according to the composition table in Figure 2.4.

A **finite group** $\mathcal{G}$ is a group with finitely many elements. For such a group the composition rule can be given by a table as in the previous example. This is called the **Cayley table** of the group $\mathcal{G}$. Only for very small groups is this practical.

The powers of an element X in a group $\mathcal{G}$ are defined by

$$X^n = \underbrace{X \cdots X}_{n \text{ times}} \qquad \text{for } n \text{ a positive integer.}$$

In particular $X^1 = X$. For $n = 0$ we set

$$X^0 = E$$

and for $m < 0$

$$X^m = (X^{-1})^{-m}.$$

With these conventions we have then the formulas

$$X^m \cdot X^n = X^{m+n}, \tag{37}$$

$$(X^m)^n = X^{m \cdot n}. \tag{38}$$

EXERCISE 2.19. Prove formulas (37), (38) by induction.

PROPOSITION 2.19. *Let $\mathcal{G}$ be a group and A, B elements of $\mathcal{G}$.*

(i) *The equation $AX = B$ has a unique solution X.*
(ii) *The equation $YA = B$ has a unique solution Y.*

Proof: (i) Multiplying on the left by A^{-1}, we have necessarily $X = A^{-1}B$, which proves uniqueness. This X is indeed a solution. (ii) Multiplying on the right by A^{-1}, we have necessarily $Y = BA^{-1}$, which proves uniqueness. This Y is indeed a solution.

PROPOSITION 2.20 (Cancellation rules). *Let $\mathcal{G}$ be a group.*

(i) *$AX = AX'$ implies $X = X'$.*
(ii) *$YA = Y'A$ implies $Y = Y'$.*

Proof: (i) Multiplying on the left by A^{-1} yields the result. (ii) is proved by multiplying by A^{-1} from the right.

For the Cayley table or **multiplication table** of a finite group $\mathcal{G}$ the cancellation rules imply the following.

PROPOSITION 2.21. *Every element of $\mathcal{G}$ occurs exactly once in every row and exactly once in every column of its multiplication table.*

An illustration of this fact in a special case is Figure 2.4.

When a subset $\mathcal{H}$ of a group $\mathcal{G}$ has the property that the composition rule in $\mathcal{G}$ turns $\mathcal{H}$ itself into a group, then $\mathcal{H}$ is called a **subgroup** of $\mathcal{G}$. $\mathcal{H}$ is a **proper** subgroup if $\mathcal{H} \neq \mathcal{G}$. The **trivial** subgroup is the group $\mathcal{H} = \{E\}$. The smallest subgroup containing a given element X is denoted $\langle X \rangle$ and is said to be **generated by** X. Such subgroups are called **cyclic**. The **order** of an element X in $\mathcal{G}$ is the least positive integer n such that $X^n = E$. If there is no such n then $\langle X \rangle$ is said to be **infinite cyclic**. Otherwise $\langle X \rangle$ is said to be **finite cyclic** of order n. The number of elements of $\langle X \rangle$ is easily seen to be precisely n. The order of a (finite) group is the number of its elements. To examine a group it is often fruitful to examine its cyclic subgroups.

EXERCISE 2.20. Prove that a subgroup $\mathcal{H}$ of $\mathcal{G}$ containing X contains the cyclic subgroup $\langle X \rangle$ generated by X.

EXERCISE 2.21. Prove that the group $\mathbf{Z}_n$ defined by the table in Figure 2.4 is cyclic of order n, and is generated by 1.

EXERCISE 2.22. Let $n = 5$. Prove that every element of $\mathbf{Z}_5$ different from 0 generates $\mathbf{Z}_5$. What is the situation for $\mathbf{Z}_6$?

EXERCISE 2.23. Compare the orders of X and X^{-1} in a group $\mathcal{G}$.

When are two groups the same? Consider Figure 2.5.

+	0	1
0	0	1
1	1	0

$\cdot$	ι	σ
ι	ι	σ
σ	σ	ι

Figure 2.5.

We have on the left the (additive) group $\mathbf{Z}_2$ and on the right the transformation group $\mathcal{G}$ of the plane generated by the central reflection σ at the origin. If 0 is matched to ι and 1 to σ, the addition on the left corresponds to the composition on the right. The two group $\mathbf{Z}_2$ and $\mathcal{G}$ are (abstractly) the same. This leads to the following concepts.

Two groups $\mathcal{G}$ and $\mathcal{G}'$ are **isomorphic**, if there is a one-to-one correspondence of the elements of $\mathcal{G}$ with all elements of $\mathcal{G}'$, such that the compositions in $\mathcal{G}$

and $\mathcal{G}'$ correspond to each other under this identification. In formulas, there is a map $f : \mathcal{G} \to \mathcal{G}'$, such that

$$f(X \cdot Y) = f(X) \cdot f(Y)$$

for $X, Y \in \mathcal{G}$. The correspondence f is an *isomorphism*. Thus $\mathbf{Z}_2$ and $\mathcal{G}$ in the situation above are isomorphic, and should be considered the same (abstract) group.

EXERCISE 2.24. Prove that any group with two elements is isomorphic to $\mathbf{Z}_2$.

EXERCISE 2.25. Prove that any group with three elements is isomorphic to $\mathbf{Z}_3$.

2.7 Symmetries of a rectangle. It is convenient to use the following notation for a permutation σ of a finite set Γ, e.g., $\Gamma = \{A, B, C, D\}$. One writes

$$\sigma = \begin{pmatrix} A & B & C & D \\ \sigma(A) & \sigma(B) & \sigma(C) & \sigma(D) \end{pmatrix}.$$

Since σ is a permutation, the second row contains each element of Γ precisely once.

We consider now the rectangle with vertices A, B, C, D and symmetry axes h, v in Figure 2.6.

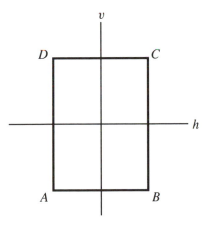

Figure 2.6.

The permutation

$$\sigma_h = \begin{pmatrix} A & B & C & D \\ D & C & B & A \end{pmatrix}$$

describes the effect of the reflection in the horizontal line h on the vertices of the rectangle. $\sigma_h^2 = \iota$ (identity) and thus σ_h is an involution. Similarly

$$\sigma_v = \begin{pmatrix} A & B & C & D \\ B & A & D & C \end{pmatrix}$$

describes the effect of the reflection in the vertical line v on the vertices of the rectangle. $\sigma_v^2 = \iota$ and σ_v is also an involution. The central reflection in the intersection point O of the lines h and v is described by

$$\sigma = \begin{pmatrix} A & B & C & D \\ C & D & A & B \end{pmatrix}.$$

The permutation σ is an involution: $\sigma^2 = \iota$. The set

$$\mathcal{V}_4 = \{\iota, \sigma, \sigma_h, \sigma_v\}$$

is a group, the Klein 4-group. It is completely described by the multiplication table in Figure 2.7.

	ι	σ	σ_h	σ_v
ι	ι	σ	σ_h	σ_v
σ	σ	ι	σ_v	σ_h
σ_h	σ_h	σ_v	ι	σ
σ_v	σ_v	σ_h	σ	ι

Figure 2.7.

We have the incomplete table in Figure 2.7′ by the preceding discussion.

	ι	σ	σ_h	σ_v
ι	ι	σ	σ_h	σ_v
σ	σ	ι		
σ_h	σ_h		ι	
σ_v	σ_v			ι

Figure 2.7′.

To complete the table to Figure 2.7, note that

$$\sigma_v \sigma_h = \begin{pmatrix} A & B & C & D \\ C & D & A & B \end{pmatrix} = \sigma_h \sigma_v = \sigma.$$

To finish the table, apply Proposition 2.21 to the second and third row to get

$$\sigma \sigma_h = \sigma_h \sigma = \sigma_v,$$

and then to the fourth and second row to get

$$\sigma_v \sigma = \sigma \sigma_v = \sigma_h.$$

EXERCISE 2.26. Why are the groups $\mathcal{V}_4$ and $\mathbf{Z}_4$ not isomorphic?

EXERCISE 2.27. Prove that every group with four elements is isomorphic to $\mathcal{V}_4$ or to $\mathbf{Z}_4$.

2.8 Symmetries of a square. If A, B, C, D denote the vertices of a square, then

$$\rho = \begin{pmatrix} A & B & C & D \\ B & C & D & A \end{pmatrix}$$

rotates the square through an angle $\pi/2$. We have

$$\rho^2 = \begin{pmatrix} A & B & C & D \\ C & D & A & B \end{pmatrix}$$

$$\rho^3 = \begin{pmatrix} A & B & C & D \\ D & A & B & C \end{pmatrix}$$

and $\rho^4 = \iota$. Thus ρ generates a cyclic group $\langle \rho \rangle$ of order 4.

But there is also the symmetry

$$\sigma = \begin{pmatrix} A & B & C & D \\ A & D & C & B \end{pmatrix}$$

of the square. $\sigma^2 = \iota$, so σ is an involution (the reflection of the square in the diagonal ℓ_{AC}). We find

$$\sigma \rho = \begin{pmatrix} A & B & C & D \\ D & C & B & A \end{pmatrix}$$

$$\sigma \rho^2 = \begin{pmatrix} A & B & C & D \\ C & B & A & D \end{pmatrix}$$

$$\sigma \rho^3 = \begin{pmatrix} A & B & C & D \\ B & A & D & C \end{pmatrix}.$$

These symmetries are all involutions. $\sigma\rho^2$ is the reflection in the diagonal ℓ_{BD}, while $\sigma\rho$ and $\sigma\rho^3$ are reflections in two other lines (which ?). We verify that

$$\sigma\rho = \rho^{-1}\sigma, \tag{39}$$

since both are given by $\begin{pmatrix} A & B & C & D \\ D & C & B & A \end{pmatrix}$. Note that $\rho^{-1} = \rho^3$, since $\rho^4 = \iota$. The relation (39) says then that $\sigma\rho = \rho^3\sigma$. In particular $\sigma\rho^3 \neq \rho^3\sigma$ and the group of symmetries of a square is not commutative.

The group $\{\iota, \rho, \rho^2, \rho^3; \sigma, \sigma\rho, \sigma\rho^2, \sigma\rho^3\}$ is denoted $\mathscr{D}_4$. It is called the **dihedral group** of order 8; the subscript 4 indicates the order of ρ. We wish to complete the multiplication table of $\mathscr{D}_4$.

PROPOSITION 2.22. *The multiplication table of $\mathscr{D}_4$ is completely determined by the relations*

$$\rho^4 = \iota, \quad \sigma^2 = \iota, \quad \sigma\rho = \rho^{-1}\sigma.$$

Proof: The last relation multiplied from the left by ρ yields $\rho\sigma\rho = \sigma$, and this relation multiplied from the right by ρ^{-1} yields by (39)

$$\rho\sigma = \sigma\rho^{-1} = \sigma\rho^3. \tag{40}$$

Thus

$$\rho^2\sigma = \rho(\rho\sigma) = \rho(\sigma\rho^{-1}) = (\rho\sigma)\rho^{-1} = (\sigma\rho^3)\rho^{-1} = \sigma\rho^2 \tag{41}$$

$$\rho^3\sigma = \rho^{-1}\sigma = \sigma\rho. \tag{42}$$

Further,

$$\rho(\sigma\rho) = (\rho\sigma)\rho = (\sigma\rho^{-1})\rho = \sigma$$
$$\rho^2(\sigma\rho) = (\rho^2\sigma)\rho = (\sigma\rho^2)\rho = \sigma\rho^3$$
$$\rho^3(\sigma\rho) = (\rho^3\sigma)\rho = (\sigma\rho)\rho = \sigma\rho^2.$$

The only missing elements in the multiplication table in Figure 2.8 are now determined by these identities and Proposition 2.21.

2.9 Symmetries of an equilateral triangle. If A, B, C denote the vertices of such a triangle, then

$$\rho = \begin{pmatrix} A & B & C \\ B & C & A \end{pmatrix}$$

$$\sigma = \begin{pmatrix} A & B & C \\ A & C & B \end{pmatrix}$$

satisfy $\rho^3 = \iota$, $\sigma^2 = \iota$ and $\sigma\rho = \rho^{-1}\sigma$. The proof of the last relation is the verification that both sides are represented by $\begin{pmatrix} A & B & C \\ C & B & A \end{pmatrix}$. Note that $\rho^2\sigma$ and $\sigma\rho^2$ are not the same.

	ι	ρ	ρ^2	ρ^3	σ	$\sigma\rho$	$\sigma\rho^2$	$\sigma\rho^3$
ι	ι	ρ	ρ^2	ρ^3	σ	$\sigma\rho$	$\sigma\rho^2$	$\sigma\rho^3$
ρ	ρ	ρ^2	ρ^3	ι	$\sigma\rho^3$	σ	$\sigma\rho$	$\sigma\rho^2$
ρ^2	ρ^2	ρ^3	ι	ρ	$\sigma\rho^2$	$\sigma\rho^3$	σ	$\sigma\rho$
ρ^3	ρ^3	ι	ρ	ρ^2	$\sigma\rho$	$\sigma\rho^2$	$\sigma\rho^3$	σ
σ	σ	$\sigma\rho$	$\sigma\rho^2$	$\sigma\rho^3$	ι	ρ	ρ^2	ρ^3
$\sigma\rho$	$\sigma\rho$	$\sigma\rho^2$	$\sigma\rho^3$	σ	ρ^3	ι	ρ	ρ^2
$\sigma\rho^2$	$\sigma\rho^2$	$\sigma\rho^3$	σ	$\sigma\rho$	ρ^2	ρ^3	ι	ρ
$\sigma\rho^3$	$\sigma\rho^3$	σ	$\sigma\rho$	$\sigma\rho^2$	ρ	ρ^2	ρ^3	ι

Figure 2.8.

PROPOSITION 2.23. *The group* $\mathcal{D}_3 = \{\iota,\ \rho,\ \rho^2;\ \sigma,\ \sigma\rho,\ \sigma\rho^2\}$ *is completely determined by the relations*

$$\rho^3 = \iota, \quad \sigma^2 = \iota, \quad \sigma\rho = \rho^{-1}\sigma. \tag{43}$$

Proof: $\sigma\rho = \rho^{-1}\sigma$ is equivalent to $\rho\sigma\rho = \sigma$ or

$$\rho\sigma = \sigma\rho^{-1}.$$

It follows that

$$\rho\sigma = \sigma\rho^2$$
$$\rho^2\sigma = \rho^{-1}\sigma = \sigma\rho$$

and

$$\rho(\sigma\rho) = (\rho\sigma)\rho = (\sigma\rho^{-1})\rho = \sigma$$
$$\rho^2(\sigma\rho) = (\rho^2\sigma)\rho = (\sigma\rho)\rho = \sigma\rho^2.$$

These relations and Proposition 2.21 completely determine the multiplication table in Figure 2.9.

	ι	ρ	ρ^2	σ	$\sigma\rho$	$\sigma\rho^2$
ι	ι	ρ	ρ^2	σ	$\sigma\rho$	$\sigma\rho^2$
ρ	ρ	ρ^2	ι	$\sigma\rho^2$	σ	$\sigma\rho$
ρ^2	ρ^2	ι	ρ	$\sigma\rho$	$\sigma\rho^2$	σ
σ	σ	$\sigma\rho$	$\sigma\rho^2$	ι	ρ	σ^2
$\sigma\rho$	$\sigma\rho$	$\sigma\rho^2$	σ	ρ^2	ι	ρ
$\sigma\rho^2$	$\sigma\rho^2$	σ	$\sigma\rho$	ρ	ρ^2	ι

Figure 2.9.

2.10 Dihedral groups. For every $n \geq 2$ the dihedral group $\mathscr{D}_n$ is the group consisting of the elements

$$\{\iota, \rho, \ldots, \rho^{n-1}; \sigma, \sigma\rho, \ldots, \sigma\rho^{n-1}\},$$

and whose multiplication table is completely determined by the relations

$$\rho^n = \iota, \quad \sigma^2 = \iota, \quad \sigma\rho = \rho^{-1}\sigma. \tag{44}$$

For $n = 2$ we have

$$\rho^2 = \iota, \quad \sigma^2 = \iota, \quad \sigma\rho = \rho\sigma$$

and $\mathscr{D}_2$ is isomorphic to the commutative Klein group $\mathscr{V}_4$. For $n > 2$ the dihedral group $\mathscr{D}_n$ is not commutative.

The subgroup $\mathscr{C}_n = \langle\rho\rangle$ of $\mathscr{D}_n$ is a cyclic subgroup of order n. The order of $\mathscr{D}_n$ is $2n$.

EXERCISE 2.28. What is the symmetry group of the Chrysler logo in Figure 2.10?

Figure 2.10.

EXERCISE 2.29. What is the symmetry group of the Chevrolet Celebrity wheel cover in Figure 2.11?

Figure 2.11.

To summarize what we have done in this chapter: we have looked at plane geometry from the point of view of groups of transformations. This is Klein's point of view in his Erlanger program: there are as many geometries as there are groups. In retrospect, the affine geometry topics of Chapter 1 are seen to be associated to the group of collineations of the plane. In the remaining chapters, we turn to Euclidean Geometry, defined by the group of motions or isometries. First we have to discuss how to measure lengths and angles.

3

Up to now we have discussed geometry in general, and in particular affine geometry. In this chapter we consider Euclidean geometry, which is in addtion based on length and angle measurements. This is most conveniently done with the help of the scalar product.

3.1 Definition and elementary properties. Let $X = (x_1, x_2)$ and $Y = (y_1, y_2)$ be two vectors. Their *scalar product* is defined by

$$X \cdot Y = x_1 y_1 + x_2 y_2.$$

The *length* of X is the positive square root

$$|X| = \sqrt{X \cdot X} = \sqrt{x_1^2 + x_2^2}.$$

The *distance* from X to Y is defined by

$$d(X, Y) = |X - Y|.$$

The scalar product has the following properties for vectors X, Y, Z and r a real number:

(SP1) $X \cdot Y = Y \cdot X$

(SP2) $(X + Y) \cdot Z = X \cdot Z + Y \cdot Z$

(SP3) $(rX) \cdot Y = r(X \cdot Y)$

(SP4) $X \cdot X \geq 0$, and $X \cdot X = 0$ iff $X = O$.

EXERCISE 3.1. Verify the properties (SP1) to (SP4).

Properties (SP1) and (SP2) imply that

$$X \cdot (Y + Z) = X \cdot Y + X \cdot Z.$$

Properties (SP1) and (SP3) imply that

$$X \cdot (rY) = r(X \cdot Y).$$

(SP2), (SP3) and these relations express the linearity of $X \cdot Y$ viewed as a function of X or a function of Y. Together they are expressed by saying that the scalar product is *bilinear*. (SP1) says the scalar product is symmetric. (SP4) is the property of *positive definiteness* of the scalar product. In the calculations below only these properties of the scalar product will be used, and not the specific formula defining it in terms of the components of X and Y.

E.g., we evaluate by these rules

$$\begin{aligned} (X + Y) \cdot (X + Y) &= X \cdot (X + Y) + Y \cdot (X + Y) \\ &= X \cdot X + X \cdot Y + Y \cdot X + Y \cdot Y \\ &= |X|^2 + 2(X \cdot Y) + |Y|^2. \end{aligned} \tag{1}$$

EXERCISE 3.2. Verify similarly that

$$(X - Y) \cdot (X - Y) = |X|^2 - 2(X \cdot Y) + |Y|^2. \tag{2}$$

Further we have by the same argument

$$(X + Y) \cdot (X - Y) = |X|^2 - |Y|^2. \tag{3}$$

Note that the scalar product is completely determined by the length function. This follows from (1), which implies

$$X \cdot Y = \frac{1}{2} \left(|X + Y|^2 - |X|^2 - |Y|^2 \right). \tag{4}$$

The identities (1) and (2) add up to

$$|X + Y|^2 + |X - Y|^2 = 2 \left(|X|^2 + |Y|^2 \right), \tag{5}$$

which is called the *parallelogram law*. The vectors $X + Y$ and $X - Y$ are diagonal vectors in the parallelogram O, X, $X + Y$, Y in Figure 3.1.

EXERCISE 3.3. Prove that if two medians of a triangle are of equal length, then the triangle is *isosceles*. The calculations are simplified substantially, if one assumes that the triangle is as in Figure 3.2 with one vertex at the origin.

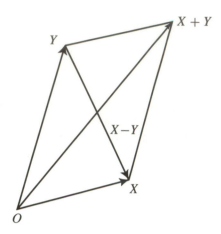

Figure 3.1.

3.2 Orthogonality. Two vectors X, Y are said to be *orthogonal* (or perpendicular), if $X \cdot Y = 0$.

E.g., $E_1 = (1, 0)$ and $E_2 = (0, 1)$ are orthogonal, since $E_1 \cdot E_2 = 1 \cdot 0 + 0 \cdot 1 = 0$. The vectors $X = (1, 1)$ and $Y = (-1, 1)$ are orthogonal, since $X \cdot Y = -1 + 1 = 0$. The vectors $X = (x_1, x_2)$ and $X' = (-x_2, x_1)$ are orthogonal, since $X \cdot X' = -x_1 x_2 + x_2 x_1 = 0$.

EXERCISE 3.4. Prove that a vector A which is orthogonal to all vectors is necessarily the vector O.

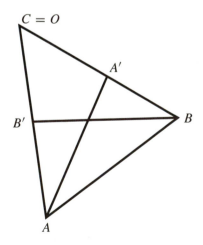

Figure 3.2.

EXERCISE 3.5. Prove that for two non-zero orthogonal vectors A and B the line ℓ_{AB} does not contain O.

A *rhombus* is defined as a parallelogram with sides of equal length.

PROPOSITION 3.1. *A parallelogram is a rhombus if and only if its diagonals are orthogonal.*

Proof: We can assume that the parallelogram is situated as in Figure 3.1 with vertices O, X, $X + Y$, Y. The vectors $X + Y$ and $X - Y$ are diagonal vectors. The identity (3) proves that the diagonal vectors are orthogonal exactly when the sides of the parallelogram have equal length.

If the bisection point of the diagonals of a parallelogram is at the origin, the vertices are A, B, $-A$, $-B$ as in Figure 3.3. The vectors $2A$, $2B$ are diagonal

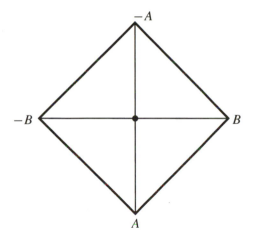

Figure 3.3.

vectors and the vectors $A - B$ and $A + B$ are sides of the parallelogram. By Proposition 3.1 the parallelogram is a rhombus exactly when $A \cdot B = 0$, which may clarify the definition of orthogonality.

The *perpendicular bisector* n of the segment from A to B ($A \neq B$) is the line n perpendicular to ℓ_{AB} through the midpoint $M = \frac{1}{2}(A + B)$ (see Figure 3.4). Consider for any X the parallelogram A, X, B, Y with $Y - A = B - X$. By Proposition 3.1 this is a rhombus if and only if ℓ_{XY} and ℓ_{AB} are orthogonal. It follows that for a point X we have

$$X \in n \quad \text{if and only if} \quad |X - A| = |X - B|. \tag{6}$$

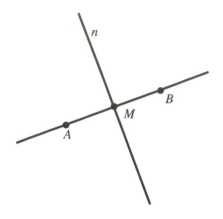

Figure 3.4.

THEOREM 3.2. *The perpendicular bisectors of the sides of a triangle are concurrent.*

The point of concurrence is called the ***circumcenter*** D of $\triangle ABC$. It is equidistant from the vertices A, B, C.

Circumcircle

"Do not despair. Remember there is no triangle, however obtuse, but the circumference of some circle passes through its wretched vertices".

Samuel Beckett, *Murphy*

EXERCISE 3.6. Prove Theorem 3.2.

THEOREM 3.3 (Theorem of Pythagoras). *The scalar product $X \cdot Y = 0$ if and only if $|Y - X|^2 = |X|^2 + |Y|^2$ (see Figure 3.5).*

Proof: By (2) we have $|Y - X|^2 = |Y|^2 - 2Y \cdot X + |X|^2$

The following facts have the same formal content (and proofs):

$$|X + Y|^2 = |X|^2 + |Y|^2 \quad \text{if and only if} \quad X \cdot Y = 0$$
$$|X - Y|^2 = |X|^2 + |Y|^2 \quad \text{if and only if} \quad X \cdot Y = 0$$
$$|X + Y|^2 = |X - Y|^2 \quad \text{if and only if} \quad X \cdot Y = 0.$$

A ***rectangle*** is a parallelogram with orthogonal sides. The last stated equivalence proves the following.

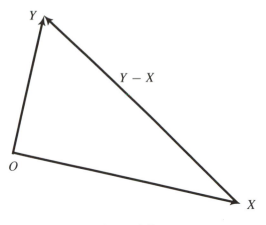

Figure 3.5.

PROPOSITION 3.4. *A parallelogram is a rectangle if and only if its diagonals are of equal length.*

The **altitude** ℓ_C of the triangle $\triangle ABC$ through the vertex C is the line ℓ_C perpendicular to ℓ_{AB} through C (see Figure 3.6). Its intersection point H_C

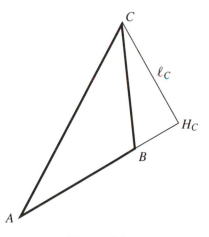

Figure 3.6.

with ℓ_{AB} is the **foot** of ℓ_C. Similarly, we obtain the altitudes ℓ_A and ℓ_B with feet H_A and H_B.

THEOREM 3.5. *The altitudes of a triangle are concurrent.*

The point of concurrence is called the **orthocenter** H of $\triangle ABC$.

First proof of Theorem 3.5: This proof is based on the concurrence of the perpendicular bisectors (Theorem 3.2) and the proof idea in Exercise 2.7 of Chapter 2. Let G be the centroid of $\triangle ABC$ and consider the dilatation $\delta = \delta_{G,-2}$. We look at Figure 2.3, but for $P = D$ (circumcenter). In this case the lines a', b', c' are the perpendicular bisectors of the sides of $\triangle ABC$. Since $\delta(A') = A$, the line a' is transformed by δ into a line through A parallel to a', i.e., the altitude ℓ_A. Thus δ transforms the perpendicular bisectors into the altitudes. Since the perpendicular bisectors are concurrent, so are the altitudes.

Note that this argument applied to the dilation $\delta_{G,-1/2}$ instead, proves conversely that the concurrence of the altitudes implies the concurrence of the perpendicular bisectors.

Second proof of Theorem 3.5: This proof is based on the identity

$$(X - A) \cdot (B - C) + (X - B) \cdot (C - A) + (X - C) \cdot (A - B) = 0 \qquad (7)$$

This identity holds for any X, and it follows from the expansion of the LHS, using the biadditivity of the scalar product.

Let now H be the intersection point of the two altitudes ℓ_A and ℓ_B. Note that

$$H \in \ell_A \Rightarrow (H - A) \cdot (B - C) = 0$$
$$H \in \ell_B \Rightarrow (H - B) \cdot (C - A) = 0$$

It follows from (7) applied to $X = H$ that $(H - C) \cdot (A - B) = 0$. But this means that $H \in \ell_C$.

Third proof of Theorem 3.5: We consider $\triangle ABC$ as the triangle of the midpoints of the sides of a triangle $\triangle A'B'C'$ as in Figure 3.7. The altitude ℓ_C of $\triangle ABC$

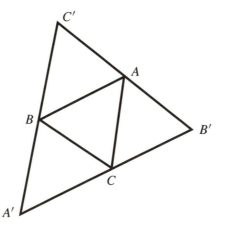

Figure 3.7.

through C is the perpendicular bisector of $\triangle A'B'C'$ through C'. Thus the orthocenter of $\triangle ABC$ and the circumcenter of $\triangle A'B'C'$ coincide.

The first proof of Theorem 3.5 given above shows that

$$H = \delta_{G,-2}(D) \tag{8}$$

for the orthocenter H and the circumcenter D. By formula (13) of Chapter 2 we have

$$H = \delta_{G,-2}(D) = G - 2(D - G) = 3G - 2D. \tag{9}$$

This implies

$$3(G - D) = H - D. \tag{10}$$

Both (8) or (10) show that G, D and H are collinear. This line is called the **Euler line** of $\triangle ABC$. There is one case where this line is not well-defined, namely when the three points coincide. This is the case for an equilateral triangle, where median, perpendicular bisector and altitude are exactly the same line. Using $G = \frac{1}{3}(A + B + C)$ (from formula (15) of Chapter 1), we find from (9)

$$H = A + B + C - 2D. \tag{11}$$

This formula is particularly simple if we choose the origin as circumcenter D. Then

$$H = A + B + C. \tag{12}$$

The following Figure 3.8 will illustrate the situation ($D = O$). Here we have

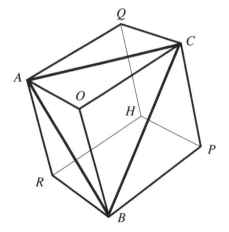

Figure 3.8.

constructed rhombi $OARB$, $OAQC$, $OBPC$. Let H be such that $AQHR$ is a rhombus. Then $PCQH$ and $PBRH$ are rhombi. Further

$$(H - A) \cdot (R - Q) = 0 \qquad \text{(diagonals in a rhombus)}$$

and $R - Q = B - C$. Thus

$$(H - A) \cdot (B - C) = 0.$$

Similarly

$$(H - B) \cdot (C - A) = 0$$
$$(H - C) \cdot (A - B) = 0$$

and H is the orthocenter (see the second proof of Theorem 3.5). Note that H clearly satisfies (12) (it may help to interpret $H - O$ as space diagonal in a parallelepiped, all projected to the plane).

It is further clear that for $\triangle PQR$ the roles of O and H are reversed: O is its orthocenter and H its circumcenter. (Make the above construction for $\triangle PQR$. You will end up with the same drawing.) Thus the Euler lines of $\triangle ABC$ and $\triangle PQR$ are the same.

Another point of the Euler line is of interest. This is the midpoint N of the interval connecting the circumcenter D and the orthocenter H. If $D = O$ as above, then by (12)

$$N = \frac{1}{2}(A + B + C). \tag{13}$$

It is the center of the *nine-point circle* discussed in the next section.

EXERCISE 3.7. For $\triangle ABC$ consider the midpoint A'' of A and H, and the midpoint B'' of B and H, where H is the orthocenter. Let A', B' be the midpoints of the sides of $\triangle ABC$ opposite A and B. Prove that $A''B''A'B'$ is a rectangle.

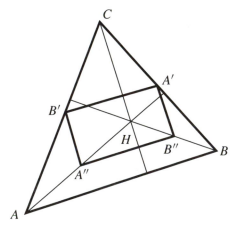

Figure 3.9.

3.3 Circles. The points of the circle with center D and radius r are characterized by

$$|X - D| = r.$$

THEOREM 3.6 (Theorem of Thales). *In* $\triangle ABC$, *let* $\mathscr{C}$ *be the circle with side* AB *as diameter. Then the angle at* C *is a right angle if and only if* C *is a point of the circle* $\mathscr{C}$ *(see Figure 3.10).*

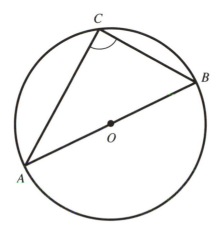

Figure 3.10.

Proof: Let the origin O be the center of $\mathscr{C}$. Then $B = -A$. Thus

$$(C - A) \cdot (C - B) = (C - A) \cdot (C + A) = |C|^2 - |A|^2.$$

The lines ℓ_{AC} and ℓ_{BC} are orthogonal if and only if the LHS equals zero. The RHS equals zero if and only if $|C| = |A|$, i.e., if only if C is a point of $\mathscr{C}$.

EXERCISE 3.8. Consider Figure 3.10 and complete it with the $\triangle ABD$ obtained from $\triangle ABC$ by central reflection in the center of the circle. State a theorem in terms of the quadrilateral $ADBC$.

EXERCISE 3.9. For two points $A \neq B$ let

$$\mathscr{C}_r = \{X \mid |X - A| = r|X - B|\} \qquad \text{for } r > 0.$$

For $r = 1$, these are precisely the points of the perpendicular bisector of ℓ_{AB}. Prove that for $r \neq 1$, $\mathscr{C}_r$ is a circle.

The following result is used in our next Theorem.

PROPOSITION 3.7. *The image of a circle under a dilatation is a circle.*

Proof: Let $\mathscr{C}$ be a circle with center D and radius r. For a point X of $\mathscr{C}$ we have then $|X - D| = r$. For a translation τ_A

$$|\tau_A(X) - \tau_A(D)| = |X - D| = r,$$

which shows that the image of $\mathscr{C}$ is the circle with center $\tau_A(D)$ and the same radius. For a central dilatation δ_s

$$|\delta_s(X) - \delta_s(D)| = |s(X - D)| = |s| \cdot |X - D| = |s| \cdot r$$

which shows that the image of $\mathscr{C}$ is the circle with center $\delta_s(D)$ and radius $|s| \cdot r$. For a central dilatation $\delta_{C,s} = \tau_C \delta_s \tau_C^{-1}$, a repeated application of these arguments shows that the image of $\mathscr{C}$ is a circle.

THEOREM 3.8   (Nine-Point Circle Theorem).   *Consider $\triangle ABC$ with orthocenter H, and the nine following points. Let A', B', C' be the midpoints of the sides opposite A, B, C. Let*

$$A'' = \frac{1}{2}(A + H), \quad B'' = \frac{1}{2}(B + H), \quad C'' = \frac{1}{2}(C + H).$$

Let D, E, F be the feet of the altitudes ℓ_A, ℓ_B, ℓ_C. Then these nine points lie on a circle. Its center N is on the Euler line, and it is the midpoint of the circumcenter and the orthocenter H of $\triangle ABC$.

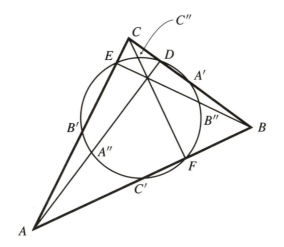

Figure 3.11.

This circle is often called the *Euler circle* or the *Feuerbach circle*.

Proof: Let G be the centroid of $\triangle ABC$. The dilatation $\delta = \delta_{G,-1/2}$ maps the circumcircle $\mathscr{C}'$ through A, B, C into a circle $\mathscr{C}$ by Proposition 3.7. Since δ maps A, B, C to A', B', C', the points A', B', C' lie on $\mathscr{C}$. It remains to show that the other six points also lie on $\mathscr{C}$.

It is convenient to assume that the circumcenter is at the origin. Then

$$\delta(O) = \delta_{G,-1/2}(O) = \frac{3}{2}G = \frac{1}{2}(A + B + C).$$

By (13) this shows

$$\delta(O) = N$$

for the prospective center of $\mathscr{C}$. It follows that N is the midpoint of O and H.

For any line ℓ the feet of the lines orthogonal to ℓ through O and H are equidistant from the midpoint N. To see this, consider also the line orthogonal to ℓ though N, and apply the Pythagorean theorem to the two resulting right triangles. It follows that

$$|A' - N| = |D - N|, \quad |B' - N| = |E - N|, \quad |C' - N| = |F - N|.$$

Since A', B', C' are on $\mathscr{C}$, so are D, E, F.

Consider the dilatation $\gamma = \delta_{H,1/2}$. Then

$$\gamma(O) = \delta_{H,1/2}(O) = \frac{1}{2}H = N.$$

Thus the image circle of $\mathscr{C}'$ under γ is a circle with center N. Its radius is one half the radius of $\mathscr{C}'$. Thus it must coincide with $\mathscr{C}$. Since γ maps A, B, C to A'', B'', C'', this completes the proof.

EXERCISE 3.10. Use Exercise 3.7 and the Theorem of Thales to give another proof of the Nine-Point Circle Theorem.

3.4 Cauchy-Schwarz inequality. This inequality states that for any two vectors X and Y

$$(X \cdot Y)^2 \le |X|^2 \cdot |Y|^2. \tag{14}$$

Moreover equality holds if and only if $X = rY$ or $Y = sX$.

Proof: For any real number t we have

$$0 \le |X + tY|^2 = |X|^2 + 2tX \cdot Y + t^2|Y|^2 = P(t).$$

If $Y = O$, (14) is an equality and $Y = 0X$. Thus we can assume $Y \neq O$, so that $P(t)$ is a quadratic polynomial in t. Since $P(t)$ is non-negative, it follows that the discriminant

$$(X \cdot Y)^2 - |X|^2 \cdot |Y|^2 \leq 0.$$

(Think of the formula for the solutions of the quadratic equation $P(t) = 0$. The condition $P(t) \geq 0$ says that there is at most one solution.) This proves (14). Moreover $P(t_0) = 0$ is realized for (exactly) one t_0 precisely when $(X \cdot Y)^2 = |X|^2 \cdot |Y|^2$. This means that $X + t_0 Y = O$ or $X = -t_0 Y$. Geometrically this means that the line ℓ_{XY} goes through the origin.

A consequence of (14) is the triangle inequality

$$|X + Y| \leq |X| + |Y|, \tag{15}$$

which, if X is replaced by $X - Y$, leads to

$$|X - Y| \geq |X| - |Y|. \tag{16}$$

Proof of (15)*:* This follows from

$$
\begin{aligned}
|X + Y|^2 &= |X|^2 + 2X \cdot Y + |Y|^2 \\
&\leq |X|^2 + 2|X| \cdot |Y| + |Y|^2 \\
&= (|X| + |Y|)^2,
\end{aligned}
$$

where we have used (14).

The proof shows that equality holds in (15) if and only if

$$X \cdot Y = |X| \cdot |Y|.$$

This is the case if and only if $X = rY$ or $Y = sX$, and necessarily $r > 0$, $s > 0$ since $|X| \cdot |Y| > 0$. It follows that one of the two vectors X, Y is between the other and O.

As noted in the beginning of this chapter, the **distance** between two points X, Y is given by

$$d(X, Y) = |X - Y| = \sqrt{(X - Y) \cdot (X - Y)} \tag{17}$$

(positive square root). The distance has the following three properties:

(D1) $d(X, Y) = d(Y, X)$

(D2) $d(X, Y) \geq 0$; $d(X, Y) = 0$ iff $X = Y$

(D3) $d(X, Z) \leq d(X, Y) + d(Y, Z)$

In (D3) equality holds if and only if, say, $Y - Z = t(X - Y)$ with $t \geq 0$, so that $(1 + t)Y = tX + Z$, which puts Y on the line segment between X and Z.

EXERCISE 3.11. Prove (D3).

EXERCISE 3.12. Let A, B, C, D be the vertices of a convex quadrilateral. Convexity means that for each of the lines ℓ_{AB}, ℓ_{BC}, ℓ_{CD}, ℓ_{DA} the quadrilateral lies in one of its half-planes. Find the point P for which the minimum

$$\text{Min}(d(P, A) + d(P, B) + d(P, C) + d(P, D))$$

is realized.

3.5 Projection. We calculate the *orthogonal projection* of a vector Y to the line defined by the nonzero vector X (Figure 3.12). The projection vector $\text{proj}_X Y$ is

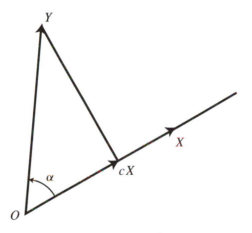

Figure 3.12.

a scalar multiple of X: $\text{proj}_X Y = cX$. The projection vector cX is characterized by the property

$$(Y - cX) \cdot X = 0,$$

or $Y \cdot X - X \cdot (cX) = 0$, which leads to the equation

$$c = \frac{X \cdot Y}{|X|^2} \tag{18}$$

and the formula

$$\text{proj}_X Y = cX = \frac{X \cdot Y}{|X|^2} \cdot X. \tag{19}$$

The length of the projection vector is

$$|cX| = |c| \cdot |X| = \frac{|(X \cdot Y)|}{|X|}. \tag{20}$$

If X is a *unit vector*, i.e., $|X| = 1$, then

$$\text{proj}_X Y = (X \cdot Y)X \tag{21}$$

and

$$|\text{proj}_X Y| = |(X \cdot Y)| \tag{22}$$

Remark. An application of the theorem of Pythagoras to the right triangle in Figure 3.12 yields

$$|cX|^2 + |Y - cX|^2 = |Y|^2.$$

Then multiplication by $|X|^2$ and formula (18) yields

$$(X \cdot Y)^2 + |X|^2|Y - cX|^2 = |X|^2|Y|^2.$$

The formula

$$|X|^2|Y - cX|^2 = |X|^2|Y|^2 - (X \cdot Y)^2$$

explains the deficiency term in the Cauchy-Schwarz inequality (14). The deficiency vanishes if and only if $Y - cX = O$, which is the condition that X and Y are on the same line through O.

EXERCISE 3.13. Let E_1 and E_2 be two orthogonal vectors of unit length. Prove that any vector X can be written in the form

$$X = (X \cdot E_1)E_1 + (X \cdot E_2)E_2.$$

EXERCISE 3.14. Let area$(\triangle OXY)$ denote the area of the triangle $\triangle OXY$ defined by the vectors X and Y. Verify the formula

$$\text{area}(\triangle OXY) = \frac{1}{2}\sqrt{|X|^2|Y|^2 - (X \cdot Y)^2}.$$

3.6 Angles. Let α be the oriented angle between X and Y, denoted $\alpha = \sphericalangle(X, Y)$. This notation means that α is positive if X turns to Y in the sense of the positive orientation of the plane, i.e., counter clockwise as in Figure 3.12. Then

$$\cos \alpha = \frac{X \cdot Y}{|X||Y|}. \tag{23}$$

Proof: From Figure 3.12 and (18)

$$\cos \alpha = \frac{c|X|}{|Y|} = \frac{X \cdot Y}{|X||Y|}.$$

This argument is certainly valid for $-\pi/2 < \alpha < \pi/2$. For the other angles formula (23) corresponds exactly to the correct signs.

The main consequence of (23) is the following fact.

COROLLARY 3.9. *The angles in plane geometry are completely determined by the scalar product (up to sign).*

Note that by formula (4) the scalar product is itself expressible in terms of length of vectors. Thus the principle above can be restated in the following form: *The angles in plane geometry are completely determined by length (up to sign).*

EXERCISE 3.15. Let area($\triangle OXY$) denote the area of the triangle $\triangle OXY$ defined by the vectors X and Y. Verify the formula

$$\text{area}(\triangle OXY) = \frac{1}{2}|X||Y| \cdot |\sin \sphericalangle(X, Y)|.$$

Remark. To determine the sign of $\sphericalangle(X, Y)$, additional information aside from (23) is needed. If $X = (x_1, x_2)$ and $Y = (y_1, y_2)$, then

$$\det(X, Y) = \begin{vmatrix} x_1 & x_2 \\ y_1 & y_2 \end{vmatrix} = x_1 y_2 - x_2 y_1$$

is positive iff $0 < \sphericalangle(X, Y) < \pi$, and negative iff $-\pi < \sphericalangle(X, Y) < 0$ (the case $\det(X, Y) = 0$ corresponds to $\sphericalangle(X, Y) = 0$ or π).

The proof of this statement is based on the fact that

$$|\det(X, Y)| = \text{area}(X, Y)$$

where area(X, Y) is the area of the parallelogram defined by X, Y. To verify this formula, observe that by Exercise 3.14 area$(X, Y) = \sqrt{|X|^2|Y|^2 - (X \cdot Y)^2}$. For $X = (x_1, x_2)$, $Y = (y_1, y_2)$ this yields the expression

$$\text{area}(X, Y) = \sqrt{(x_1^2 + x_2^2)(y_1^2 + y_2^2) - (x_1 y_1 + x_2 y_2)^2}$$
$$= \sqrt{(x_1 y_2 - x_2 y_1)^2}$$

which coincides with $\det(X, Y)$ up to sign.

Another consequence of (23) is the law of cosines, which states that for non zero vectors X and Y

$$|X - Y|^2 = |X|^2 + |Y|^2 - 2 \cos \sphericalangle (X, Y) \cdot |X| \, |Y|. \qquad (24)$$

Proof: This follows from the expansion

$$|X - Y|^2 = |X|^2 + |Y|^2 - 2X \cdot Y$$

and (23).

Note that for orthogonal vectors X and Y this is the Theorem of Pythagoras.

EXERCISE 3.16. The area of the triangle $\triangle OXY$ defined by the vectors X and Y is given by ***Heron's formula***:

$$\text{area}(\triangle OXY) = \sqrt{s(s - a)(s - b)(s - c)}$$

where $a = |X|$, $b = |Y|$, $c = |Y - X|$ are the lengths of the sides of $\triangle OXY$, and

$$s = \frac{1}{2}(a + b + c).$$

3.7 Equation of a line. What is the equation of a line ℓ through a point P and orthogonal to a given vector $N \neq O$, as in Figure 3.13? $X \in \ell$ if and only if

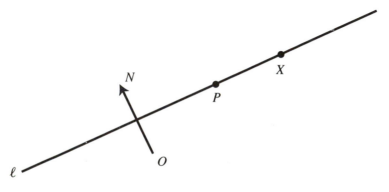

Figure 3.13.

$$(X - P) \cdot N = 0 \qquad (25)$$

This is equivalently expressed by

$$X \cdot N = P \cdot N \qquad (26)$$

EXERCISE 3.17. Let $X = (x_1, x_2)$, $P = (p_1, p_2)$, and $N = (n_1, n_2)$. Write out (26) in a familiar form.

EXERCISE 3.18. Let ℓ be orthogonal to N, ℓ' orthogonal to N'. What is the condition for ℓ and ℓ' to be orthogonal?

To sum up this chapter: we discussed length and angle measurements as defined via a scalar product. In the next chapter, we return to the transformation point of view of Chapter 2, as it applies to Euclidean Geometry.

4

We consider now transformations preserving length and angle measurements. These are called isometries. Euclidean geometry studies properties invariant under isometries. The main result of this chapter is a description of the group of isometries in terms of reflections. The chapter concludes with a discussion of all possible finite group of isometries.

4.1 Definition and examples. An *isometry* is a distance preserving map. Thus a map α of the plane to itself is an isometry if

$$d(\alpha(X), \alpha(Y)) = d(X, Y) \qquad \text{for all } X, Y.$$

PROPOSITION 4.1. *Translations are isometries.*

Proof: Let τ_A be a translation. Then

$$d(\tau_A(X), \tau_A(Y)) = |\tau_A(X) - \tau_A(Y)|$$
$$= |(A + X) - (A + Y)| = |X - Y| = d(X, Y).$$

For a central dilatation δ_r we have

$$d(\delta_r(X), \delta_r(Y)) = |\delta_r(X) - \delta_r(Y)| = |rX - rY| = |r||X - Y| = |r|d(X, Y).$$

Thus distances get multiplied (dilatated by) the factor $|r|$, and δ_r is not an isometry, unless $r = \pm 1$ (in which case the dilatation is the identity map or the central reflection in the origin).

PROPOSITION 4.2. *The composition of two isometries is an isometry.*

Proof: Let α and β be isometries. Then for any X and Y

$$d(X, Y) = d(\alpha(X), \alpha(Y)) = d(\beta(\alpha(X)), \beta(\alpha(Y))),$$

which proves that $\beta\alpha$ is an isometry.

We will prove a little bit later that an isometry has an inverse (Theorem 4.6), that this inverse is also an isometry (Theorem 4.7), and that as a consequence the isometries form a group (Theorem 4.8). We prepare for this by the following considerations.

For an isometry α and $A = \alpha(O)$ we consider $\beta = \tau_A^{-1}\alpha$. By Proposition 4.2 this is an isometry. Further,

$$\beta(O) = (\tau_A^{-1}\alpha)(O) = \tau_A^{-1}(A) = A - A = O.$$

Thus any isometry α can be written in the form

$$\alpha = \tau_A\beta, \tag{1}$$

where β is an isometry satisfying $\beta(O) = O$.

THEOREM 4.3. *Let α be an isometry satisfying $\alpha(O) = O$. Then α has the following properties:*

(i) $\alpha(X + Y) = \alpha(X) + \alpha(Y)$ (additivity),

(ii) $\alpha(rX) = r\alpha(X)$ (homogeneity).

The properties (i) and (ii) together are expressed by saying that α is a ***linear map***.

Note that a linear map α necessarily satisfies $\alpha(O) = \alpha(X - X) = \alpha(X) - \alpha(X) = O$. So the statement is that the linearity of an isometry α is characterized by $\alpha(O) = O$.

Proof: First we prove that an isometry α satisfying $\alpha(O) = O$ is length preserving, i.e.,

$$|\alpha(X)| = |X| \qquad \text{for all } X. \tag{2}$$

This follows from

$$|\alpha(X)| = |\alpha(X) - \alpha(O)| = d(\alpha(X), \alpha(O)) = d(X, O) = |X - O| = |X|.$$

Next we prove that α preserves scalar products, i.e.,

$$\alpha(X) \cdot \alpha(Y) = X \cdot Y \qquad \text{for all } X, Y. \tag{3}$$

This follows by comparing the expansion

$$d(\alpha(X), \alpha(Y))^2 = |\alpha(X) - \alpha(Y)|^2 = |\alpha(X)|^2 - 2\alpha(X) \cdot \alpha(Y) + |\alpha(Y)|^2$$

with

$$d(X, Y)^2 = |X - Y|^2 = |X|^2 - 2X \cdot Y + |Y|^2$$

and applying (2).

Now we turn to the proof of the additivity property (i). Using (2) and (3) we evaluate

$$
\begin{aligned}
|\alpha(X + Y) &- (\alpha(X) + \alpha(Y))|^2 \\
&= |\alpha(X + Y)|^2 - 2\alpha(X + Y) \cdot (\alpha(X) + \alpha(Y)) + |\alpha(X) + \alpha(Y)|^2 \\
&= |\alpha(X + Y)|^2 - 2\alpha(X + Y) \cdot \alpha(X) - 2\alpha(X + Y) \cdot \alpha(Y) \\
&\quad + |\alpha(X)|^2 + 2\alpha(X) \cdot \alpha(Y) + |\alpha(Y)|^2 \\
&= |X + Y|^2 - 2(X + Y) \cdot X - 2(X + Y) \cdot Y + |X|^2 + 2X \cdot Y + |Y|^2 \\
&= |X + Y|^2 - 2(X + Y) \cdot (X + Y) + |X + Y|^2 \\
&= 0.
\end{aligned}
$$

This implies (i).

To prove the homogeneity property (ii), we similarly evaluate

$$
\begin{aligned}
|\alpha(rX) - r\alpha(X)|^2 &= |\alpha(rX)|^2 - 2r\,\alpha(rX) \cdot \alpha(X) + r^2|\alpha(X)|^2 \\
&= |rX|^2 - 2r(rX \cdot X) + r^2|X|^2 \\
&= 0,
\end{aligned}
$$

which implies (ii).

EXERCISE 4.1. Prove that any isometry α is angle preserving up to sign, i.e.,

$$\sphericalangle(\alpha(X), \alpha(Y)) = \pm\sphericalangle(X, Y).$$

THEOREM 4.4. *An isometry maps lines into lines.*

As stated earlier, a map with this property is called a **collineation**. Translations and central dilatations are examples of collineations.

Proof: First we consider the case of a linear isometry. Then the identity

$$\alpha(aA + bB) = a\alpha(A) + b\alpha(B)$$

proves that the points of the line ℓ_{AB} are mapped by α to the points of the line $\ell_{\alpha(A)\alpha(B)}$.

For the case of a translation τ_A the desired property has already been established (Proposition 2.4). An arbitrary isometry α is by (1) a linear isometry followed by a translation. Therefore α maps lines into lines.

COROLLARY 4.5. *An isometry maps parallel lines into parallel lines.*

Proof: This follows from Theorem 4.4 and the fact that an isometry preserves angles up to sign (Exercise 4.1).

THEOREM 4.6. *An isometry is a bijection.*

Proof: To prove that an isometry α is one-to-one, assume $\alpha(X) = \alpha(Y)$. Then

$$0 = d(\alpha(X), \alpha(Y)) = d(X, Y),$$

which implies $X = Y$.

It remains to prove that α is a map *onto* the plane. Since translations are bijections, it suffices to prove this for the case of a linear isometry. Let m and n be a pair of orthogonal lines through the origin. Then $\alpha(m)$ and $\alpha(n)$ are also a pair of orthogonal lines through O. Given nonzero vectors X, Y on lines m, n, respectively, we get vectors $\alpha(X), \alpha(Y)$ on lines $\alpha(m), \alpha(n)$, respectively. Now let Q be any point of the plane. Then there are numbers a and b such that

$$Q = a\alpha(X) + b\alpha(Y).$$

Define

$$P = aX + bY.$$

Then $\alpha(P) = Q$.

By Theorem 4.6 every isometry has an inverse.

PROPOSITION 4.7. *The inverse of an isometry is an isometry.*

Proof: For the inverse α^{-1} of an isometry we have

$$d(\alpha^{-1}(X), \alpha^{-1}(Y)) = d(\alpha(\alpha^{-1}(X)), \alpha(\alpha^{-1}(Y))) = d(X, Y),$$

i.e., α^{-1} is also distance preserving.

We can summarize Propositions 4.2 and 4.7 in the following statement.

THEOREM 4.8. *The set of all isometries is a group.*

This is the isometry group of the plane, which according to Klein's Erlanger Program (see Chapter 2) characterizes Euclidean geometry. It is also called the group of motions or congruences of the Euclidean plane. Two figures in the plane are congruent, when they are equivalent under an isometry. This is the precise meaning of the conclusion of the so-called congruence theorems of plane geometry.

4.2 Fixed points of isometries. Recall that a fixed point of a bijection α (or more generally a map $\alpha: \mathcal{V} \to \mathcal{V}$) is a point X such that $\alpha(X) = X$. For a linear isometry, e.g., the origin is a fixed point. A translation with a fixed point is the identity (Proposition 2.3). The following fact will prove to be important.

PROPOSITION 4.9. *Let α be an isometry. If X and Y are fixed points of α, then every point P of the line ℓ_{XY} is a fixed point of α.*

Proof: Since $\alpha(X) = X$ and $\alpha(Y) = Y$, the line ℓ_{XY} is mapped to itself by α. Thus for $P \in \ell_{XY}$ necessarily $\alpha(P) \in \ell_{XY}$. Since α is distance preserving, $d(\alpha(P), X) = d(P, X)$ and $d(\alpha(P), Y) = d(P, Y)$, and it follows that $\alpha(P) = P$.

Another proof can be based on the fact α can be written as a linear isometry β followed by a translation τ_A, $\alpha = \tau_A \beta$, or equivalently $\beta = \tau_A^{-1} \alpha$. Then

$$\beta(X) = (\tau_A^{-1}\alpha)(X) = \alpha(X) - A = X - A$$
$$\beta(Y) = (\tau_A^{-1}\alpha)(Y) = \alpha(Y) - A = Y - A.$$

Any $P \in \ell_{XY}$ is of the form

$$P = aX + bY \qquad \text{with } a + b = 1.$$

It follows that

$$\begin{aligned}
\beta(P) = \beta(aX + bY) &= a\beta(X) + b\beta(Y) \\
&= a(X - A) + b(Y - A) = aX + bY - (a + b)A \\
&= P - A
\end{aligned}$$

or equivalently $\alpha(P) = (\tau_A \beta)(P) = P$.

PROPOSITION 4.10. *Let α be an isometry. If α has three fixed points which are not collinear, then $\alpha = \iota$ (identity).*

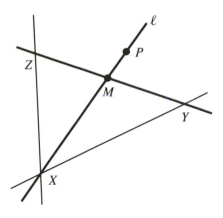

Figure 4.1.

Proof: Let X, Y and Z be three non-collinear fixed points. Then every point of the lines ℓ_{XY}, ℓ_{YZ} and ℓ_{XZ} is a fixed point. Let P be any point different from X, and M the intersection point of ℓ_{YZ} with $\ell = \ell_{XP}$ (see Figure 4.1). Since Y and Z are fixed points, so is every point of ℓ_{YZ} (Proposition 4.9). Thus M is a fixed point, and so every point of ℓ is also. Thus P is fixed. For points of the line through X parallel to ℓ_{YZ} the same argument applies if we replace the line ℓ_{XP} by ℓ_{YP}.

EXERCISE 4.2. Let α and β be isometries, and A, B, C three non-collinear points for which $\alpha(A) = \beta(A), \alpha(B) = \beta(B)$ and $\alpha(C) = \beta(C)$. Prove that $\alpha = \beta$.

4.3 Reflections. Let us first consider the following geometric definition of the *reflection* σ_ℓ *in a line* ℓ. It is the bijection of the plane which leaves every point of ℓ fixed and maps every point $P \notin \ell$ to the point $P' = \sigma_\ell(P)$ such that the line ℓ is the perpendicular bisector of the line segment between P and P' (see Figure 4.2). σ_ℓ is an isometry and from this description, $\sigma_\ell^2 = \iota$, i.e., σ_ℓ is an involution.

To find a formula for σ_ℓ expressing these facts, first assume ℓ to be a line through the origin and Y a unit vector on ℓ (see Figure 4.3). The diagonal on ℓ of the rhombus in Figure 4.3 is twice the orthogonal projection $\text{proj}_Y X$ of X to ℓ. But by formula (21) of Chapter 3 we have

$$\text{proj}_Y X = (X \cdot Y)Y.$$

Now ℓ is the perpendicular bisector of X and $\sigma_\ell(X)$ at say Q. Thus

$$\text{proj}_Y X = (X \cdot Y)Y = Q = \frac{X + \sigma_\ell(X)}{2}.$$

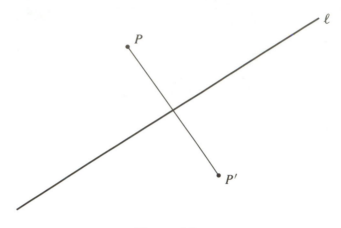

Figure 4.2.

Thus

$$\sigma_\ell(X) = -X + 2(X \cdot Y)Y. \qquad (4)$$

This formula can serve as an analytic definition of the reflection in the line ℓ through the origin. Note that for $X \in \ell$ we have $(X \cdot Y)Y = X$ and by (4) therefore $\sigma_\ell(X) = -X + 2X = X$.

An alternative formula for the reflection σ_ℓ in a line through O is obtained by considering the vector

$$U = \operatorname{proj}_Y X - X \qquad (5)$$

from X to the orthogonal projection of X to Y. Then σ_ℓ is characterized by

$$\sigma_\ell(X) - X = 2U$$

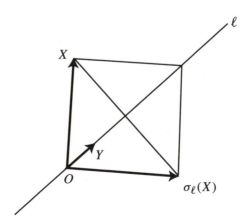

Figure 4.3.

or

$$\sigma_\ell(X) = X + 2U. \tag{6}$$

EXERCISE 4.3. Verify from these formulas alone that σ_ℓ is an isometry.

EXERCISE 4.4. Verify from formula (4) that σ_ℓ is an involution.

The following fact is central to our discussion.

THEOREM 4.11. *Let α be an isometry. If α has two distinct fixed points P and Q, then either α is the reflection σ_ℓ in the line $\ell = \ell_{PQ}$ or $\alpha = \iota$ (identity).*

Proof: Assume $\alpha \neq \iota$. Then there exists a point R such that $R' = \alpha(R) \neq R$. By Proposition 4.9 the point R is not on the line $\ell = \ell_{PQ}$. Then

$$d(P, R) = d(\alpha(P), \alpha(R)) = d(P, R'),$$

so that P is a point of the perpendicular bisector of R and R'. Similarly

$$d(Q, R) = d(\alpha(Q), \alpha(R)) = d(Q, R')$$

shows that Q is a point of the same perpendicular bisector. It follows that $\ell = \ell_{PQ}$ is the perpendicular bisector of R and R'. Thus $\alpha = \sigma_\ell$.

COROLLARY 4.12. *Let α be an isometry. Assume that α is an involution and leaves every point of the line ℓ fixed. Then α is the reflection σ_ℓ in ℓ.*

Proof: Since α is not the identity, α can have no further fixed point outside ℓ (Proposition 4.10). Thus $\alpha = \sigma_\ell$ by Theorem 4.11.

The following discussion illustrates the important concept of "conjugation by isometries." Let σ_ℓ be a reflection and α an isometry. The conjugate $\alpha\sigma_\ell\alpha^{-1}$ of σ_ℓ by α is an isometry. It is described as follows.

PROPOSITION 4.13. $\alpha\sigma_\ell\alpha^{-1} = \sigma_{\alpha(\ell)}$, *i.e., the conjugate of σ_ℓ by an isometry α is the reflection in the line $\alpha(\ell)$.*

Proof: By Corollary 4.12 is suffices to verify two facts, namely (i) that $\alpha \sigma_\ell \alpha^{-1}$ is an involution and (ii) that $\alpha \sigma_\ell \alpha^{-1}$ leaves every point of the line $\alpha(\ell)$ fixed.

(i) is verified by the calculation

$$(\alpha \sigma_\ell \alpha^{-1})^2 = \alpha \sigma_\ell^2 \alpha^{-1} = \alpha \alpha^{-1} = \iota,$$

and the observation that $\alpha \sigma_\ell \alpha^{-1} = \iota$ would imply $\alpha \sigma_\ell = \alpha$ and $\sigma_\ell = \iota$.

(ii) is verified for $Q \in \alpha(\ell)$ by

$$(\alpha \sigma_\ell \alpha^{-1})(Q) = \alpha \sigma_\ell (\alpha^{-1}(Q)) = \alpha(\alpha^{-1}(Q)) = Q,$$

since $\alpha^{-1}(Q) \in \ell$ and hence $\sigma_\ell(\alpha^{-1}(Q)) = \alpha^{-1}(Q)$.

Let α be an isometry taking a point P of ℓ to O. The conjugation formula $\alpha \sigma_\ell \alpha^{-1} = \sigma_{\alpha(\ell)}$ then shows that calculations involving reflections can be reduced to calculations involving reflections in lines through the origin, where the formulas (4) and (6) apply. Thus these formulas suffice for an analytic treatment.

EXERCISE 4.5. Prove that for an isometry α the identity $\alpha \sigma_\ell = \sigma_\ell \alpha$ for all lines ℓ implies $\alpha = \iota$.

4.4 Central reflections. We have discussed these maps already in Section 2.3. The central reflection σ_C in a point C is given by

$$\sigma_C(X) = -X + 2C \tag{7}$$

which expresses the property that C is the midpoint of X and $\sigma_C(X)$.

PROPOSITION 4.14. σ_C *is an isometry.*

Proof: This follows from

$$d(\sigma_C(X), \sigma_C(Y)) = |\sigma_C(X) - \sigma_C(Y)| = |(-X + 2C) - (-Y + 2C)|$$
$$= |Y - X| = d(X, Y).$$

We have already shown that σ_C is an involution (Proposition 2.13).

Recall further that C is the unique fixed point of σ_C (Proposition 2.14). In analogy to Corollary 4.12 we have the following fact.

THEOREM 4.15. *Let α be an isometry. Assume that α is an involution and has C as unique fixed point. Then α is the central reflection σ_C in C.*

First we prove the following auxiliary result.

LEMMA 4.16. *Let α be an involutive isometry. Then for any X the midpoint of the line segment from X to $\alpha(X)$ is a fixed point of α.*

Proof: We can assume that X is not already a fixed point, because otherwise there is nothing to prove. For the midpoint M of the segment from X to $\alpha(X)$ we have then $d(M, X) = d(M, \alpha(X))$ and hence

$$d(\alpha(M), \alpha(X)) = d(\alpha(M), \alpha^2(X)) = d(\alpha(M), X),$$

which shows that $\alpha(M)$ is a point of the perpendicular bisector of X and $\alpha(X)$. Since α maps the line $\ell = \ell_{X\alpha(X)}$ to itself, $\alpha(M)$ is a point of ℓ. Therefore $\alpha(M) = M$.

Proof of Theorem 4.15: Let X be any point distinct from C, so that $\alpha(X) = X'$ is different from X. The midpoint of the segment from X to X' is a fixed point of α by Lemma 4.16. Therefore this midpoint is C. This shows that C is the midpoint of the segment from X to $\alpha(X)$ for every $X \neq C$. Therefore $\alpha = \sigma_C$.

In analogy to Proposition 4.13 we have the following conjugation property.

PROPOSITION 4.17. *Let α be any isometry. Then $\alpha\sigma_C\alpha^{-1} = \sigma_{\alpha(C)}$, i.e., the conjugate of σ_C by α is the central reflection in $\alpha(C)$.*

Proof: By Theorem 4.15 we have to verify two facts, namely (i) that $\alpha\sigma_C\alpha^{-1}$ is an involution and (ii) that $\alpha(C)$ is the unique fixed point of $\alpha\sigma_C\alpha^{-1}$.

(i) is verified by the calculation

$$(\alpha\sigma_C\alpha^{-1})^2 = \alpha\sigma_C^2\alpha^{-1} = \alpha\alpha^{-1} = \iota,$$

and the observation that $\alpha\sigma_C\alpha^{-1} = \iota$ would imply $\alpha\sigma_C = \alpha$ and $\sigma_C = \iota$.

(ii) is verified by

$$(\alpha\sigma_C\alpha^{-1})(\alpha(C)) = (\alpha\sigma_C)(C) = \alpha(C),$$

and the fact that $(\alpha\sigma_C\alpha^{-1})(X) = X$ implies

$$(\sigma_C\alpha^{-1})(X) = \alpha^{-1}(X),$$

and hence $\alpha^{-1}(X) = C$, i.e., $X = \alpha(C)$.

EXERCISE 4.6. Prove that for an isometry α the identity $\alpha\sigma_C = \sigma_C\alpha$ for all points C implies $\alpha = \iota$.

4.5 Isometries with a unique fixed point. A central reflection has a unique fixed point. The following result is of central importance.

THEOREM 4.18. *Let α be an isometry. If α has exactly one fixed point P, then α is the product $\alpha = \sigma_n \sigma_m$ of two reflections in lines m and n intersecting in P.*

Proof: For any point X different from P, $\alpha(X) = X' \neq X$. We claim that P is on the perpendicular bisector n of X and X'. This follows from

$$d(X, P) = d(\alpha(X), \alpha(P)) = d(X', P).$$

It follows that $\sigma_n(X') = X$ and $\sigma_n(P) = P$. Thus

$$(\sigma_n \alpha)(X) = X \quad \text{and} \quad (\sigma_n \alpha)(P) = P.$$

By Theorem 4.11 an isometry with two fixed points is either a reflection or the identity. If $\sigma_n \alpha = \iota$, then $\alpha = \sigma_n$, which contradicts the assumption that α has a unique fixed point. It follows that $\sigma_n \alpha$ is the reflection in the line $m = \ell_{XP}$, i.e., $\sigma_n \alpha = \sigma_m$. Thus $\alpha = \sigma_n \sigma_m$.

The line $m = \ell_{XP}$ intersects the line n in P. This completes the proof of the Theorem.

What is the assertion of the Theorem in the case of a central reflection σ_P? In this case P is the midpoint of X and X' (for any X) so the proof leads to two perpendicular lines m, n through P, and a representation of σ_P as the product $\sigma_n \sigma_m$. Note that any pair of perpendicular lines through P will do, so the assertion in Theorem 4.18 in no way implies uniqueness of the lines m and n.

EXERCISE 4.7. Prove that the product of two reflections in perpendicular lines is the central reflection in their intersection point.

COROLLARY 4.19. *Let α be an isometry. If α has at least one fixed point, then α is the product of at most two reflections.*

Proof: If α has exactly one fixed point, then α is a product of two reflections. If α has more than one fixed point, then by Theorem 4.11 the map α is either a reflection or the identity.

THEOREM 4.20. *Every isometry can be written as the product of not more than three reflections.*

Proof: Assume α to be different from the identity, and let $\alpha(P) = Q$ for some $P \neq Q$. Let m be the perpendicular bisector of P and Q. Then

$$(\sigma_m \alpha)(P) = \sigma_m(Q) = P$$

and P is a fixed point of $\sigma_m \alpha$. By Corollary 4.11 the isometry $\sigma_m \alpha$ is the product of at most two reflections. Thus α is the product of at most three reflections.

Note that this does not preclude that an isometry might simultaneously be the product of a larger number of reflections. The assertion is that α can be written as a product involving at most three reflections.

4.6 Products of involutions. As a preliminary consideration, let $\mathcal{G}$ be any group. An involution $X \in \mathcal{G}$ is an element $X \neq E$ satisfying $X^2 = E$ (identity). Let X and Y be involutions in $\mathcal{G}$. When is XY again an involution? The answer is given by

$$(XY)^2 = XYXY = E \iff XY = YX.$$

Note that $XY = E$ implies $X = Y$, so that distinct involutions have an involution as product precisely when they commute.

A first example is the product of two commuting reflections.

PROPOSITION 4.21. *The reflections in two distinct lines m and n commute if and only if the lines are perpendicular.*

Proof: The commuting relation is

$$\sigma_m \sigma_n = \sigma_n \sigma_m. \tag{8}$$

Formula (8) is equivalent to

$$\sigma_m \sigma_n \sigma_m = \sigma_n. \tag{9}$$

By the conjugation formula in Proposition 4.13, the LHS equals the reflection in the line $\sigma_m(n)$. Thus (9) is equivalent to $\sigma_m(n) = n$. If m and n are parallel and distinct, $\sigma_m(n) \neq n$. Thus m and n intersect. But then $\sigma_m(n) = n$ if and only if m and n are perpendicular.

PROPOSITION 4.22. *The product of two commuting reflections (in perpendicular lines m and n) is the central reflection σ_P in the intersection point P of the lines m and n.*

This has been proved before (see Exercise 4.7).

When do two central reflections commute? We observe:

$$\sigma_P \sigma_Q = \sigma_Q \sigma_P$$
$$\Longleftrightarrow \sigma_Q \sigma_P \sigma_Q = \sigma_P$$
$$\Longleftrightarrow \sigma_{\sigma_Q(P)} = \sigma_P.$$

Here we have used the conjugation formula in Proposition 4.17. The last equality holds iff $\sigma_Q(P) = P$, which implies $P = Q$. Thus two distinct central reflections never commute.

Finally we consider the product of a central reflection σ_P and a reflection σ_ℓ.

PROPOSITION 4.23. *A central reflection σ_P and a reflection σ_ℓ commute iff $P \in \ell$.*

Proof: The commuting property

$$\sigma_P \sigma_\ell = \sigma_\ell \sigma_P$$

is equivalent to

$$\sigma_\ell \sigma_P \sigma_\ell = \sigma_P,$$

which by the conjugation formula in Proposition 4.17 is equivalent to

$$\sigma_{\sigma_\ell(P)} = \sigma_P \quad \text{or} \quad \sigma_\ell(P) = P.$$

But the points of the line ℓ are the only fixed points of σ_ℓ.

EXERCISE 4.8. Let $P \in \ell$. Describe the isometry $\sigma_P \sigma_\ell$ (which equals $\sigma_\ell \sigma_P$).

EXERCISE 4.9. Prove that ℓ is the perpendicular bisector of AB if and only if

$$\sigma_\ell \sigma_B \sigma_\ell \sigma_A = \iota.$$

EXERCISE 4.10. Prove that ℓ is a line parallel to the line ℓ_{AB} through A and B if and only if

$$\sigma_B \sigma_\ell \sigma_B \sigma_A \sigma_\ell \sigma_A = \iota.$$

4.7 Translations. Every isometry can be written as the product of not more than three reflections. What about translations? The answer is that two reflections suffice.

To see this, consider two parallel lines m and n. To calculate $\sigma_n \sigma_m$, let ℓ be a line perpendicular to m and n, intersecting m in M and n in N (see Figure 4.4).

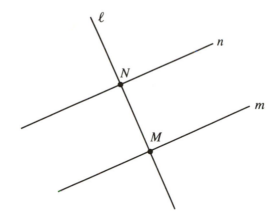

Figure 4.4.

By Exercise 4.7,

$$\sigma_N = \sigma_n \sigma_\ell \quad \text{and} \quad \sigma_M = \sigma_\ell \sigma_m.$$

It follows that

$$\sigma_n \sigma_m = \sigma_N \sigma_M. \tag{10}$$

By Theorem 2.15 we have

$$\sigma_N \sigma_M = \tau_A \quad \text{with } A = 2(N - M). \tag{11}$$

We have proved the following.

THEOREM 4.24. *The product of two reflections in parallel lines m and n is a translation:*

$$\sigma_n \sigma_m = \tau_A \quad \text{with } A = 2(N - M). \tag{12}$$

Note that the line n is given by

$$n = \tau_{A/2}(m). \tag{13}$$

EXERCISE 4.11. Prove that for two parallel lines m and n we have $\sigma_n \sigma_m = \sigma_m \sigma_n$ if and only if $m = n$.

We can clearly reverse the arguments above: beginning with a translation τ_A, we can find lines m and n such that (12) holds, and points M and N such that (11) holds. This proves the following facts.

THEOREM 4.25. *(i) A translation is the product of two reflections in parallel lines. (ii) A translation is the product of two central reflections.*

Moreover, for a given translation τ_A and a line m perpendicular to A, there is a unique line n parallel to m such that $\tau_A = \sigma_n \sigma_m$. The line n is given by (13). Similarly, for a given translation τ_A and a point M, there is a unique point N such that $\tau_A = \sigma_N \sigma_M$. The point N is given by $N = M + A/2$.

A fact used later is the following consequence of the arguments in this section.

THEOREM 4.26. *Let m, n and p be parallel lines. Then $\sigma_n \sigma_m \sigma_p$ is a reflection σ_q in a fourth line q parallel to m, n and p.*

Proof: For the translation $\sigma_n \sigma_m = \tau_A$ there is a unique line q parallel p such that $\tau_A = \sigma_q \sigma_p$. But $\sigma_n \sigma_m = \sigma_q \sigma_p$ implies $\sigma_n \sigma_m \sigma_p = \sigma_q$ as claimed.

EXERCISE 4.12. Let τ_A be a translation and P a point. Prove that $\tau_A \sigma_P$ is a central reflection with center $P + 1/2 \cdot$ A. Similarly $\sigma_P \tau_A$ is a central reflection with center $P - 1/2 \cdot A$.

EXERCISE 4.13. Let τ_A be a translation and α an isometry. Write τ_A as a product of central reflections and calculate $\alpha \tau_A \alpha^{-1}$. As a consequence, show that if α is a linear isometry, then

$$\alpha \tau_A \alpha^{-1} = \tau_{\alpha(A)}.$$

EXERCISE 4.14. Consider triangle $\triangle ABC$ and an arbitrary fourth point G. Prove that G is the centroid of $\triangle ABC$ if and only if

$$\sigma_G \sigma_C \sigma_G \sigma_B \sigma_G \sigma_A = \iota.$$

EXERCISE 4.15. Generalize Exercise 4.14 to n points $A_1, \ldots, A_n$ as follows: G is the centroid of $A_1, \ldots, A_n$ if and only if

$$(\sigma_G \sigma_{A_n})(\sigma_G \sigma_{A_{n-1}}) \cdots (\sigma_G \sigma_{A_1}) = \iota.$$

EXERCISE 4.16. What is the corresponding characterization of the centroid G of the mass-points $(a_1, A_1), \ldots, (a_n, A_n)$?

4.8 Rotations. A *rotation* with center C through the oriented angle θ is the transformation $\rho_{C,\theta}$ with fixed point C which maps a point $X \neq C$ to the point $\rho_{C,\theta}(X) = X'$ such that

$$d(C, X) = d(C, X') \quad \text{and} \quad \sphericalangle(X - C, X' - C) = \theta$$

(see Figure 4.5).

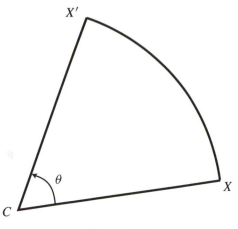

Figure 4.5.

THEOREM 4.27. *A rotation is an isometry.*

Proof: Let $X' = \rho_{C,\theta}(X)$ and $Y' = \rho_{C,\theta}(Y)$ (see Figure 4.6). We have to show

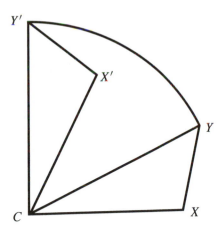

Figure 4.6.

$d(X', Y') = d(X, Y)$, or equivalently $|X' - Y'|^2 = |X - Y|^2$. But by the law of cosines (formula (24) of Chapter 3) applied to $\triangle C X' Y'$ we have

$$|X' - Y'|^2 = |X' - C|^2 + |Y' - C|^2 - 2 \cos \angle(X' - C, Y' - C)|X' - C||Y' - C|.$$

Since $|X' - C| = |X - C|$, $|Y' - C| = |Y - C|$ and

$$\angle(X' - C, Y' - C) = \angle(X - C, Y - C),$$

it follows that the corresponding expression for ΔCXY yields the same value for $|X - Y|^2$. Thus $\rho_{C,\theta}$ is distance preserving.

Let m and n be lines intersecting in a point C. Then the (oriented) angle $\measuredangle(m, n)$ from m to n is the angle between the vectors $X - C$ on m and $Y - C$ on n. For given intersecting lines there are two (oriented) angles α_1 and α_2 from m to n. But note that $\alpha_2 - \alpha_1 = \pi$, so that $2\alpha_2$ and $2\alpha_1$ coincide up to 2π.

THEOREM 4.28. *Let m and n be lines through C such that $\theta = 2\measuredangle(m, n)$. Then*

$$\rho_{C,\theta} = \sigma_n \sigma_m. \tag{14}$$

Proof: Consider a circle $\mathscr{C}$ with center C. Let M and N be the intersections of m and n with $\mathscr{C}$ (see Figure 4.7). Let $N' = \sigma_m(N)$ and $M' = \sigma_n(M)$. Then

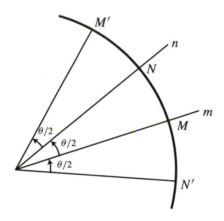

Figure 4.7.

$$\measuredangle(N' - C, M - C) = \measuredangle(M - C, N - C) = \frac{1}{2}\theta$$

and

$$\measuredangle(M - C, N - C) = \measuredangle(N - C, M' - C) = \frac{1}{2}\theta.$$

It follows that

$$\rho_{C,\theta}(M) = M' \quad \text{and} \quad \rho_{C,\theta}(N') = N.$$

On the other hand

$$(\sigma_n \sigma_m)(M) = \sigma_n(M) = M' \quad \text{and} \quad (\sigma_n \sigma_m)(N') = \sigma_n(N) = N.$$

Since both $\rho_{C,\theta}$ and $\sigma_n\sigma_m$ have C as fixed point, we find that the two isometries $\rho_{C,\theta}$ and $\sigma_n\sigma_m$ give the same result when applied to the three non-collinear points C, M and N'. By Exercise 4.2 it follows that $\rho_{C,\theta} = \sigma_n\sigma_m$.

Since for the two possible choices α_1 and α_2 of angles from m to n we have $\alpha_2 - \alpha_1 = \pi$, it follows that the corresponding rotations $\rho_{C,2\alpha_1}$ and $\rho_{C,2\alpha_2}$ coincide.

It is worthwhile comparing Theorem 4.25, representing a translation as the product of reflections in parallel lines, and Theorem 4.28, representing a rotation as the product of reflections in intersecting lines. Heuristically the statements are similar, if one thinks of a translation as a rotation with center at infinity.

For perpendicular lines m and n intersecting in C, the rotation $\sigma_n\sigma_m$ is the rotation with center C through the angle π, which is precisely the central reflection σ_C.

Another consequence of (14) is that for a given rotation $\rho_{C,\theta}$ and line m through C there is a unique line n through C such that $\rho_{C,\theta} = \sigma_n\sigma_m$. This leads to the following result in analogy to Theorem 4.26.

THEOREM 4.29. *Let m, n and p be concurrent lines (intersecting in C). Then $\sigma_n\sigma_m\sigma_p$ is a reflection in a fourth line q through C.*

Proof: For the rotation $\rho_{C,\theta} = \sigma_n\sigma_m$ and the given line p through C, there is a unique line q through C such that $\rho_{C,\theta} = \sigma_q\sigma_p$. But $\sigma_n\sigma_m = \sigma_q\sigma_p$ implies $\sigma_n\sigma_m\sigma_p = \sigma_q$ as claimed.

We recall Theorem 4.18, according to which an isometry with exactly one fixed point C is the product $\sigma_n\sigma_m$ of reflections in lines m and n intersecting in C. Applying Theorem 4.28, we obtain the following fact.

THEOREM 4.30. *An isometry with exactly one fixed point C is a rotation with center C.*

We further note that by Corollary 4.19 an isometry with at least one fixed point, is the product of at most two reflections. This implies the following fact.

COROLLARY 4.31. *An isometry distinct from the identity with at least one fixed point is either a rotation (one fixed point), or a reflection (more than one fixed point).*

The following result confirms our intuition.

THEOREM 4.32. *A rotation* $\rho_{C,\theta}$ *distinct from the identity and transforming a line* ℓ *into itself is the central reflection* σ_C. *The line* ℓ *is necessarily a line through* C.

Proof: Let m be the line perpendicular to ℓ through C. Then $\rho_{C,\theta} = \sigma_n \sigma_m$ for a unique line n through C. It follows that

$$\ell = \rho_{C,\theta}(\ell) = (\sigma_n \sigma_m)(\ell) = \sigma_n(\ell).$$

If n and ℓ have at least a common point, then $\sigma_n(\ell) = \ell$ implies either $n = \ell$ or that n and ℓ are perpendicular. In the latter case, it follows that $n = m$. But this implies $\rho_{C,\theta} = \iota$, contrary to our assumption. Thus n and ℓ are necessarily parallel (including the possibility $n = \ell$). But $\sigma_n(\ell) = \ell$ implies that $n = \ell$ and ℓ is, in particular, a line through C. Moreover

$$\rho_{C,\theta} = \sigma_n \sigma_m = \sigma_\ell \sigma_m = \sigma_C$$

as claimed.

Finally we discuss the composition of rotations. If both have center C, then clearly

$$\rho_{C,\beta} \rho_{C,\alpha} = \rho_{C,\alpha+\beta}.$$

Now consider rotations with different centers. What type of isometry is the composition?

Let $\rho_{A,2\alpha}$ and $\rho_{B,2\beta}$ be two such rotations. Let $\ell = \ell_{AB}$ and consider lines m with $\measuredangle(m, \ell) = \alpha$ and n with $\measuredangle(\ell, n) = \beta$ (note the order, see Figure 4.8).

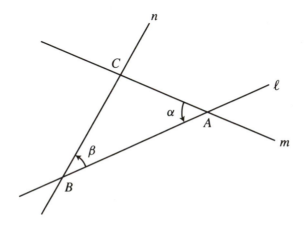

Figure 4.8.

Consider first the case where the lines m and n intersect in C. Then

$$\rho_{A,2\alpha} = \sigma_\ell \sigma_m, \qquad \rho_{B,2\beta} = \sigma_n \sigma_\ell$$

and

$$\rho_{B,2\beta} \rho_{A,2\alpha} = \sigma_n \sigma_m = \rho_{C,2\gamma} \tag{15}$$

where $\gamma = \sphericalangle(m,n) = \alpha + \beta (\neq \pi$ for intersecting lines m and n).

The other case to consider is where m and n are parallel. This is the case precisely when $\alpha + \beta = \pi$. In this case

$$\rho_{B,2\beta} \rho_{A,2\alpha} = \sigma_n \sigma_m = \tau_D \tag{16}$$

is a translation, where the vector D is given as in Theorem 4.24. This discussion can be summarized as follows.

THEOREM 4.33. *The product $\rho_{B,2\beta} \rho_{A,2\alpha}$ is either a rotation (if $\alpha + \beta \neq \pi$) or a translation (if $\alpha + \beta = \pi$). In the first case one obtains a rotation $\rho_{C,2\gamma}$, where $\gamma = \alpha + \beta$ and C is given by the construction in Figure 4.8.*

EXERCISE 4.17. Consider a triangle $\triangle ABC$ (oriented counter-clockwise) with positive angles α, β, γ at A, B, C. Prove that $\rho_{A,2\alpha} \rho_{B,2\beta} \rho_{C,2\gamma} = \iota$. Are there other similar formulas?

EXERCISE 4.18. Construct over each side of $\triangle ABC$ an equilateral triangle. The centroids G_1, G_2 and G_3 of these triangles form a new triangle, the so-called Napoleon triangle. Prove that $\triangle G_1 G_2 G_3$ is equilateral. (This theorem is attributed to Napoleon Bonaparte.)

EXERCISE 4.19. Prove that nontrivial rotations with distinct centers never commute.

4.9 Glide reflections. A *glide reflection* in the direction of a nonzero vector T is a product $\tau_T \sigma_\ell$, where ℓ is a line parallel to T.

PROPOSITION 4.34. *Let T be a nonzero vector and ℓ a line. Then ℓ is parallel to T if and only if $\tau_T \sigma_\ell = \sigma_\ell \tau_T$.*

Proof: The line ℓ is parallel to T precisely when $\tau_T(\ell) = \ell$. This can be written as

$$\sigma_{\tau_T(\ell)} = \sigma_\ell,$$

or, by the conjugation formula in Proposition 4.13,

$$\tau_T \sigma_\ell \tau_T^{-1} = \sigma_\ell,$$

or still equivalently,

$$\tau_T \sigma_\ell = \sigma_\ell \tau_T.$$

The case $T = 0$ corresponds to a reflection (no glide).

The square of a glide reflection is, by Proposition 4.34,

$$(\tau_T \sigma_\ell)^2 = \tau_T \sigma_\ell \sigma_\ell \tau_T = \tau_{2T}. \tag{17}$$

COROLLARY 4.35. *A glide reflection has no fixed point.*

Proof: If P is a fixed point of the glide reflection $\tau_T \sigma_\ell$, then it is also a fixed point of its square. Since the square is a nontrivial translation, this is impossible.

4.10 Classification of isometries. The main result of this Chapter is the following classification of isometries.

THEOREM 4.36. *Every isometry is either a reflection, or a translation, or a rotation, or a glide reflection.*

The key to this classification is Theorem 4.20, which establishes that every isometry can be written as the product of not more than 3 reflections. The product of two reflections is either a translation (if the lines are parallel) or a rotation (if the lines intersect). Therefore it remains to examine the product of three reflections in lines ℓ, m and n. If the three lines are parallel, then by Theorem 4.26 the product $\sigma_n \sigma_m \sigma_\ell$ is a reflection (in a fourth parallel line). If the three lines are concurrent, then by Theorem 4.29 the product $\sigma_n \sigma_m \sigma_\ell$ is a reflection (in a fourth concurrent line). The classification in Theorem 4.36 is therefore completely established by the following result.

THEOREM 4.37. *The product of three reflections in lines ℓ, m, n which are neither parallel nor concurrent is a glide reflection.*

Proof: The pairs ℓ and m, m and n cannot both be pairs of parallel lines. We can therefore assume, e.g., that m and n intersect in P. The third line ℓ is then not a line through P. Let $\theta = 2\sphericalangle(m, n)$. Then

$$\sigma_n \sigma_m = \rho_{P,\theta}.$$

Let s be the line perpendicular to ℓ through P (see Figure 4.9). There exists

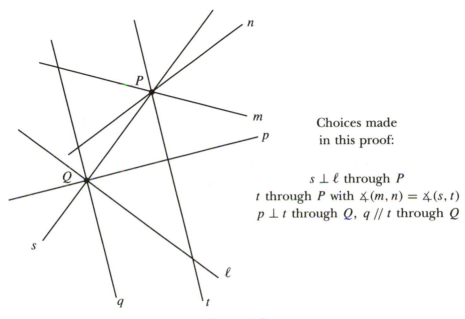

Choices made
in this proof:

$s \perp \ell$ through P
t through P with $\sphericalangle(m, n) = \sphericalangle(s, t)$
$p \perp t$ through Q, $q \,/\!/\, t$ through Q

Figure 4.9.

then a unique line t through P such that

$$\sigma_n \sigma_m = \sigma_t \sigma_s.$$

This line is characterized by $\sphericalangle(m, n) = \sphericalangle(s, t)$. We obtain then

$$\sigma_n \sigma_m \sigma_\ell = \sigma_t \sigma_s \sigma_\ell = \sigma_t \sigma_Q,$$

where Q is the intersection point of the perpendicular lines s and ℓ. Let p be the line perpendicular to t through Q and q the line parallel to t through Q. Then $\sigma_Q = \sigma_q \sigma_p$ and

$$\sigma_n \sigma_m \sigma_\ell = \sigma_t \sigma_Q = \sigma_t \sigma_q \sigma_p = \tau_T \sigma_p, \tag{18}$$

where T is perpendicular to t and q, hence parallel to p. Moreover $T = 0$ would imply $t = q$, hence $Q = P$, which contradicts the assumption that the lines ℓ, m, n are not concurrent. The formula (18) shows that $\sigma_n \sigma_m \sigma_\ell$ is indeed a glide reflection.

To show that $\sigma_\ell \sigma_n \sigma_m$ is similarly a glide reflection, one can adjust the argument above by considering the unique line t' such that

$$\sigma_n \sigma_m = \sigma_s \sigma_{t'}$$

characterized by $\sphericalangle(m, n) = \sphericalangle(t', s)$. With $p' \perp t'$ and $q' \mathbin{/\!/} t'$, we obtain then

$$\sigma_\ell \sigma_n \sigma_m = \sigma_\ell \sigma_s \sigma_{t'} = \sigma_Q \sigma_{t'} = \sigma_{p'} \sigma_{q'} \sigma_{t'} = \sigma_{p'} \tau_{T'},$$

where T' is perpendicular to q' and t', hence parallel to p'. Moreover $T' \neq 0$. Hence $\sigma_\ell \sigma_n \sigma_m$ is also a glide reflection.

The case when m and n (but not ℓ) are parallel can be reduced to these two cases by relabeling the lines appropriately.

The basic remark for the following considerations is that a reflection (in a line) changes the orientation of the plane. It follows that the product of an even number of reflections (in lines) preserves the orientation of the plane, while the product of an odd number of reflections reverses the orientation of the plane. The classification Theorem 4.36 can therefore be reformulated as follows.

THEOREM 4.38. *Every orientation preserving isometry is either a translation or a rotation. Every orientation reversing isometry is either a reflection or a glide reflection.*

COROLLARY 4.39. *The orientation preserving isometries form a subgroup of the group of isometries.*

This is not the case for the orientation reversing isometries.

Another aspect of the classification in Theorem 4.38 is that the two possibilities in each case can be distinguished by the presence or absence of fixed points.

COROLLARY 4.40. *An orientation preserving isometry without fixed points is a translation. An orientation preserving isometry with fixed points must have a unique fixed point and is a rotation. An orientation reversing isometry without fixed points must be a glide reflection. An orientation reversing isometry with fixed points must be a reflection.*

EXERCISE 4.20. Prove that the perpendicular bisectors ℓ, m and n of the sides of a triangle $\triangle ABC$ are concurrent by examining the reflections σ_ℓ, σ_m and σ_n.

EXERCISE 4.21. Consider a triangle $\triangle ABC$ (A, B, C oriented counter-clockwise) with (positive) angles α, β, γ at A, B, C. Prove that $\rho_{C,2\gamma}\rho_{B,2\beta}\rho_{A,2\alpha}$ is a translation different from the identity. (Compare the order of composition with the one in Exercise 4.17).

The following fact is of interest.

PROPOSITION 4.41. *Let $\rho_{C,\theta}$ be a rotation and α any isometry. Then*

$$\alpha\rho_{C,\theta}\alpha^{-1} = \rho_{\alpha(C),\pm\theta}.$$

If α is orientation preserving, the plus sign applies. If α is orientation reversing, the minus sign applies.

Proof: Let first $\alpha = \sigma_\ell$ be a reflection. Write $\rho_{C,\theta} = \sigma_n\sigma_m$ as a product of reflections in lines m, n intersecting in C. For a choice of m, the line n is uniquely determined by the condition $2\angle(m, n) = \theta$. It follows that

$$\sigma_\ell\rho_{C,\theta}\sigma_\ell^{-1} = \sigma_\ell\sigma_n\sigma_m\sigma_\ell = \sigma_{\sigma_\ell(n)}\sigma_{\sigma_\ell(m)}.$$

The lines $\sigma_\ell(m)$ and $\sigma_\ell(n)$ intersect in $\sigma_\ell(C)$ at angle $-\angle(m, n)$. It follows that

$$\sigma_\ell\rho_{C,\theta}\sigma_\ell = \rho_{\sigma_\ell(C),\pm\theta}.$$

If α is any isometry, we write it as a product of reflections. Repeated application of the argument just made completes the proof. The sign of the angle depends on the parity of the number of reflections used.

For a central reflection $\sigma_C = \rho_{C,\pi}$ this conjugation formula agrees with the formula found in Proposition 4.17, since $\rho_{C,\pi} = \rho_{C,-\pi}$.

For a translation τ_C, the conjugation formula in Theorem 4.41 shows

$$\tau_C\rho_\theta\tau_C^{-1} = \rho_{C,\theta}, \tag{19}$$

where ρ_θ is the rotation through angle θ about the origin.

COROLLARY 4.42. *Any rotation can be written in the form*

$$\rho_{C,\theta} = \tau_A\rho_\theta, \qquad \text{where } A = C - \rho_\theta(C).$$

Proof: For any X we have by (19)

$$\rho_{C,\theta}(X) = (\tau_C \rho_\theta \tau_C^{-1})(X) = C + \rho_\theta(X - C) = \rho_\theta(X) + C - \rho_\theta(C),$$

where we have used the linearity of ρ_θ.

For the conjugate of a translation or a glide reflection, we obtain the following special result.

PROPOSITION 4.43. *Let α be a linear isometry. Then for a vector T and a line ℓ*

$$\alpha \tau_T \alpha^{-1} = \tau_{\alpha(T)} \quad \text{and} \quad \alpha \tau_T \sigma_\ell \alpha^{-1} = \tau_{\alpha(T)} \sigma_{\alpha(\ell)}.$$

Proof: The first of these formulas is proved as in Exercise 4.13. Using this we have then

$$\alpha \tau_T \sigma_\ell \alpha^{-1} = (\alpha \tau_T \alpha^{-1})(\alpha \sigma_\ell \alpha^{-1}) = \tau_{\alpha(T)} \sigma_{\alpha(\ell)}.$$

By (1), an arbitrary isometry α is of the form $\alpha = \tau_A \beta$, where β is a linear isometry and $A = \alpha(O)$. It follows by Proposition 4.43 that

$$\alpha \tau_T \alpha^{-1} = (\tau_A \beta)\tau_T(\tau_A \beta)^{-1} = \tau_A(\beta \tau_T \beta^{-1})\tau_A^{-1} = \tau_A \tau_{\beta(T)} \tau_A^{-1} = \tau_{\beta(T)}$$

and similarly,

$$\alpha \tau_T \sigma_\ell \alpha^{-1} = (\alpha \tau_T \alpha^{-1})(\alpha \sigma_\ell \alpha^{-1}) = \tau_{\beta(T)} \sigma_{\alpha(\ell)},$$

which shows precisely to which extent the conjugation formulas proved before depended on the linearity of α.

4.11 Finite groups of isometries. We are going to determine all possible finite subgroups $\mathcal{G}$ of the group of isometries. The first observation is the following result.

THEOREM 4.44. *A finite group of isometries contains only reflections and rotations.*

Proof: The reason is that translations and glide reflections generate infinite cyclic groups. This is a consequence of formula (17).

We have seen such finite groups of isometries in Chapter 2, namely the cyclic groups $\mathcal{C}_n$ and the dihedral groups $\mathcal{D}_n$. The main result of this section is as follows.

THEOREM 4.45 (Leonardo da Vinci). *A finite group of isometries is either cyclic (if it contains only rotations) or a dihedral group (if it also contains reflections).*

The attribution of Theorem 4.45 to Leonardo da Vinci originates in H. Weyl's comments in his book "Symmetry" (see Bibliography). He notes that Leonardo engaged in systematically determining the possible symmetries of a central building, and how to attach chapels and niches without destroying the symmetries of the nucleus. This is recorded in one of Leonardo's notebooks, the repositories of his ideas.

COROLLARY 4.46. *If $\mathcal{G}$ is a finite group of isometries containing at least one reflection, then the order of $\mathcal{G}$ is even, say 2n, and $\mathcal{G}$ contains exactly n rotations (including ι), and exactly n reflections.*

The first step in the proof is the following result.

THEOREM 4.47. *All the rotations in a finite group of isometries have the same center.*

Proof: Let $\rho_{A,\alpha}$ and $\rho_{B,\beta}$ be two nontrivial rotations in a finite group $\mathcal{G}$ of isometries. The composition is not a translation, hence $\alpha + \beta \neq 2\pi$. Since the angle of

$$\gamma = \rho_{B,\beta}^{-1} \rho_{A,\alpha}^{-1} \rho_{B,\beta} \rho_{A,\alpha}$$

is zero and γ cannot be a translation, it follows that $\gamma = \iota$. This means that the two rotations commute. By Exercise 4.19 the centers A and B are therefore the same.

The second step in the proof of Leonardo's Theorem is the following special case.

THEOREM 4.48. *A finite group of isometries containing only rotations is cyclic.*

Proof: The group $\mathcal{G}$ consists of rotations with the same center A through angles α satisfying $0 \leq \alpha < 2\pi$. Let θ be the smallest occurring positive angle and $\rho = \rho_{A,\theta}$. Since the cyclic subgroup $\langle \rho \rangle$ generated by ρ is a subgroup of $\mathcal{G}$, $\langle \rho \rangle$ is finite. Let n be the order of $\langle \rho \rangle$. Then $\rho^n = \iota$ and $\theta = 2\pi/n$. We claim that $\langle \rho \rangle = \mathcal{G}$.

For let $\varphi \in \mathcal{G}$, but $\varphi \notin \langle \rho \rangle$. The angle of the rotation φ is between $i\theta$ and $(i + 1)\theta$ for some $i = 0, \ldots, n - 1$. It follows that $\rho^{-i}\varphi$ is a rotation through an angle between 0 and θ, contradicting the fact that θ is the smallest positive angle for rotations in $\mathcal{G}$. Thus no such φ exists and $\langle \rho \rangle = \mathcal{G}$.

Proof of Theorem 4.45: Let $\mathcal{G}$ be a finite group of isometries containing at least one reflection $\sigma = \sigma_\ell$. It remains to prove that $\mathcal{G}$ is a dihedral group.

Let $\mathcal{H} \subset \mathcal{G}$ be the subgroup of all rotations in $\mathcal{G}$. By Theorem 4.48 the subgroup $\mathcal{H}$ is cyclic, say of order n, i.e., $\mathcal{H} = \langle \rho \rangle$ with $\rho^n = \iota$. The case $\mathcal{H} = \{\iota\}$ corresponds to $n = 1$.

Consider $\alpha \in \mathcal{G}$, but $\alpha \notin \mathcal{H}$. Then α is a reflection. Now $\sigma \alpha$ is not a translation, hence it is a rotation, i.e., $\sigma \alpha = \rho^m$ for some $m = 0, 1, \ldots, n - 1$. It follows that $\alpha = \sigma \rho^m$. To show that $\mathcal{G}$ is isomorphic to $\mathcal{D}_n$, it only remains to verify $\sigma \rho = \rho^{-1} \sigma$, since by line (44) in Chapter 2 this relation, together with $\rho^n = \iota$, $\sigma^2 = \iota$, completely determines $\mathcal{D}_n$. But $\sigma \rho$ reverses orientation, hence is a reflection, and therefore $(\sigma \rho)^2 = \iota$. This shows $\sigma \rho = \rho^{-1} \sigma$ and completes the proof of Leonardo's Theorem.

5

LINEAR MAPS AND MATRICES

In this chapter we explain how linear maps are represented by matrices. Composition of linear maps then corresponds to matrix multiplication. We then define and examine the trace and the determinant of a linear map. As an application, we prove the crystallographic restriction theorem on the possible orders for a rotation in a so-called wallpaper symmetry group.

5.1 The matrix of a linear map. We repeat here the definition of a linear map $\alpha \colon \mathcal{V} \to \mathcal{V}$. It is characterized by the properties

$$\alpha(X + Y) = \alpha(X) + \alpha(Y) \tag{1}$$

$$\alpha(rX) = r\alpha(X) \tag{2}$$

for vectors X, Y and scalars r. As a consequence, $\alpha(0) = \alpha(X - X) = \alpha(X) - \alpha(X) = 0$. The special case discussed in Chapter 4 is when α is an isometry.

Let $E_1 = (1, 0)$ and $E_2 = (0, 1)$, the standard basis vectors in the plane. Then every vector $X = (x_1, x_2)$ has a standard representation

$$X = x_1 E_1 + x_2 E_2 \tag{3}$$

and by (1) and (2),

$$\alpha X = x_1 \alpha(E_1) + x_2 \alpha(E_2). \tag{4}$$

This equation says that a linear map α is completely determined by its effect on the standard basis vectors E_1 and E_2. We define four numbers α_{ij} $(i, j = 1, 2)$ by

$$\alpha(E_1) = \alpha_{11} E_1 + \alpha_{21} E_2$$
$$\alpha(E_2) = \alpha_{12} E_1 + \alpha_{22} E_2. \tag{5}$$

The (2×2) *matrix of* α with respect to the standard basis is the array

$$\begin{pmatrix} \alpha_{11} & \alpha_{12} \\ \alpha_{21} & \alpha_{22} \end{pmatrix} = (\alpha_{ij}).$$

The number i is the row index and j the column index. Thus the first column is the representation of $\alpha(E_1)$ in terms of the standard basis, and the second column the corresponding representation of $\alpha(E_2)$. We say in short that $\alpha(E_j)$ is the jth column vector of the matrix of α.

As an illustration we calculate $\alpha(X)$ for $X = (x_1, x_2)$ from the matrix of α:

$$\alpha(X) = \alpha(x_1 E_1 + x_2 E_2) = x_1(\alpha_{11} E_1 + \alpha_{21} E_2) + x_2(\alpha_{12} E_1 + \alpha_{22} E_2)$$
$$= (x_1 \alpha_{11} + x_2 \alpha_{12}) E_1 + (x_1 \alpha_{21} + x_2 \alpha_{22}) E_2 = x_1' E_1 + x_2' E_2$$

where

$$\begin{aligned} x_1' &= \alpha_{11} x_1 + \alpha_{12} x_2 \\ x_2' &= \alpha_{21} x_1 + \alpha_{22} x_2. \end{aligned} \tag{6}$$

This is symbolically written

$$\begin{pmatrix} x_1' \\ x_2' \end{pmatrix} = \begin{pmatrix} \alpha_{11} & \alpha_{12} \\ \alpha_{21} & \alpha_{22} \end{pmatrix} \begin{pmatrix} x_1 \\ x_2 \end{pmatrix}. \tag{7}$$

The matrix (α_{ij}) acts on the vector $\begin{pmatrix} x_1 \\ x_2 \end{pmatrix}$ and yields the vector $\begin{pmatrix} x_1' \\ x_2' \end{pmatrix}$, where x_1' and x_2' are given by (6). If $X = (x_1, x_2)$ and $X' = (x_1', x_2')$, then (7) is equivalent to the fact that

$$X' = \alpha(X).$$

Conversely we see that given any matrix (α_{ij}), the formulas (6) or (7) define a linear map α by $\alpha(X) = X'$. Thus there is a bijective correspondence between linear maps of $\mathbf{R}^2$ and (2×2) matrices.

Examples. Let $\alpha = \sigma_h$ denote the reflection in the x-axis. Since $\sigma_h(E_1) = E_1$ and $\sigma_h(E_2) = -E_2$, the matrix associated to σ_h is

$$\begin{pmatrix} 1 & 0 \\ 0 & -1 \end{pmatrix}.$$

Similarly, for the reflection σ_v in the y-axis, we have $\sigma_v(E_1) = -E_1$ and $\sigma_v(E_2) = E_2$. Therefore the matrix associated to σ_v is

$$\begin{pmatrix} -1 & 0 \\ 0 & 1 \end{pmatrix}.$$

The matrix associated to the identity map ι is the matrix

$$\begin{pmatrix} 1 & 0 \\ 0 & 1 \end{pmatrix} = I.$$

It is called the *identity matrix*.

The *zero matrix*

$$\begin{pmatrix} 0 & 0 \\ 0 & 0 \end{pmatrix} = 0$$

is associated to the map given by $\alpha(X) = 0$ for all X, the **zero map**.

The first three examples were all isometries, the last one of course is not.

Which is the linear map α represented by the matrix

$$\begin{pmatrix} 1 & 0 \\ 0 & 0 \end{pmatrix}?$$

The conditions are $\alpha(E_1) = E_1$ and $\alpha(E_2) = 0$, and α is the orthogonal projection of the plane onto the x-axis. Similarly

$$\begin{pmatrix} 0 & 0 \\ 0 & 1 \end{pmatrix}$$

represents the orthogonal projection of the plane onto the y-axis.

A further example is the central dilatation δ_r with the origin as center, which is represented by the matrix

$$\begin{pmatrix} r & 0 \\ 0 & r \end{pmatrix}.$$

EXERCISE 5.1. Let $A = (a_1, a_2)$ be a unit vector and ℓ the line through the origin in direction A. Determine the matrix of the orthogonal projection of the plane onto the line ℓ. Express it in terms of the angle between the x-axis and ℓ.

5.2 Composition of maps and matrix multiplication. If $\alpha: \mathcal{V} \to \mathcal{V}$ and $\beta: \mathcal{V} \to \mathcal{V}$ are linear maps, the composition $\gamma = \beta\alpha$ is again a linear map.

EXERCISE 5.2. Verify the linearity properties for the composition of linear maps.

Let α be represented by the matrix $A = (\alpha_{ij})$ and β by the matrix $B = (\beta_{ij})$. We wish to calculate the matrix $C = (\gamma_{ij})$ associated to the composition $\gamma = \beta\alpha$.

We denote $X' = \alpha(X)$ and $X'' = \beta(X')$, so that $X'' = \gamma(X)$. Let

$$X = (x_1, x_2), \quad X' = (x_1', x_2'), \quad X'' = (x_1'', x_2'').$$

Then by (6)

$$x_1' = \alpha_{11}x_1 + \alpha_{12}x_2$$
$$x_2' = \alpha_{21}x_1 + \alpha_{22}x_2$$

and similarly,

$$x_1'' = \beta_{11}x_1' + \beta_{12}x_2' = \beta_{11}(\alpha_{11}x_1 + \alpha_{12}x_2) + \beta_{12}(\alpha_{21}x_1 + \alpha_{22}x_2)$$
$$= (\beta_{11}\alpha_{11} + \beta_{12}\alpha_{21})x_1 + (\beta_{11}\alpha_{12} + \beta_{12}\alpha_{22})x_2$$
$$x_2'' = \beta_{21}x_1' + \beta_{22}x_2' = \beta_{21}(\alpha_{11}x_1 + \alpha_{12}x_2) + \beta_{22}(\alpha_{21}x_1 + \alpha_{22}x_2)$$
$$= (\beta_{21}\alpha_{11} + \beta_{22}\alpha_{21})x_1 + (\beta_{21}\alpha_{12} + \beta_{22}\alpha_{22})x_2.$$

This can be written

$$x_1'' = \gamma_{11}x_1 + \gamma_{12}x_2$$
$$x_2'' = \gamma_{21}x_1 + \gamma_{22}x_2,$$

with

$$\gamma_{11} = \beta_{11}\alpha_{11} + \beta_{12}\alpha_{21}$$
$$\gamma_{12} = \beta_{11}\alpha_{12} + \beta_{12}\alpha_{22}$$
$$\gamma_{21} = \beta_{21}\alpha_{11} + \beta_{22}\alpha_{21}$$
$$\gamma_{22} = \beta_{21}\alpha_{12} + \beta_{22}\alpha_{22}.$$

These formulas can be written in a unified way as

$$\gamma_{ik} = \beta_{i1}\alpha_{1k} + \beta_{i2}\alpha_{2k} \qquad (i, k = 1, 2),$$

or even more succinctly

$$\gamma_{ik} = \sum_{j=1}^{2} \beta_{ij}\alpha_{jk}. \tag{8}$$

Note that the summation index for the matrix B is the column index, while for the matrix A it is the row index.

THEOREM 5.1. *Let α, β be linear maps with associated matrices $A = (\alpha_{ij})$, $B = (\beta_{ij})$. Then the matrix $C = (\gamma_{ij})$ associated to the composition $\gamma = \beta\alpha$ is given by (8). This formula defines the product $C = BA$ of the matrices A and B.*

Thus the *multiplication of matrices* is defined as the matrix associated to the composition of the corresponding linear maps. This proves, e.g., that the multiplication of matrices is associative, i.e.,

$$C(BA) = (CB)A$$

since this is (even tautologically) the case for the composition of linear maps.

EXERCISE 5.3. Verify that the identity matrix I satisfies $AI = IA = A$ for all matrices A.

EXERCISE 5.4. Calculate the product

$$\begin{pmatrix} -1 & 0 \\ 0 & 1 \end{pmatrix} \begin{pmatrix} 1 & 0 \\ 0 & -1 \end{pmatrix}$$

and interpret the result.

EXERCISE 5.5. Calculate the products

$$\begin{pmatrix} 1 & 0 \\ 0 & -1 \end{pmatrix} \begin{pmatrix} 1 & 0 \\ 0 & -1 \end{pmatrix} \quad \text{and} \quad \begin{pmatrix} -1 & 0 \\ 0 & 1 \end{pmatrix} \begin{pmatrix} -1 & 0 \\ 0 & 1 \end{pmatrix}.$$

EXERCISE 5.6. Calculate A^2, B^2 for

$$A = \begin{pmatrix} 0 & 0 \\ a & 0 \end{pmatrix}, \qquad B = \begin{pmatrix} b & 0 \\ 0 & 0 \end{pmatrix}.$$

5.3 Matrices for linear isometries. A linear isometry is either a rotation ρ_θ with the origin as center, or a reflection σ_ℓ in a line ℓ through the origin. This is a consequence of the classification Theorem 4.36 or the intermediate result Corollary 4.31.

The matrix associated to ρ_θ is determined by the equations

$$\rho_\theta(E_1) = \cos\theta \cdot E_1 + \sin\theta \cdot E_2$$
$$\rho_\theta(E_2) = -\sin\theta \cdot E_1 + \cos\theta \cdot E_2.$$

The corresponding matrix is

$$\begin{pmatrix} \cos\theta & -\sin\theta \\ \sin\theta & \cos\theta \end{pmatrix}.$$

EXERCISE 5.7. Calculate the product of the matrices associated to the rotations ρ_{θ_1} and ρ_{θ_2}.

The matrix associated to the reflection σ_ℓ in a line ℓ through the origin can be determined as follows. If σ_h denotes as earlier the reflection in the x-axis, then the product $\sigma_\ell \sigma_h$ is the rotation $\rho_{2\theta}$, where θ is the angle between the x-axis and the line ℓ. It follows that

$$\sigma_\ell = \rho_{2\theta}\sigma_h = \begin{pmatrix} \cos 2\theta & -\sin 2\theta \\ \sin 2\theta & \cos 2\theta \end{pmatrix} \begin{pmatrix} 1 & 0 \\ 0 & -1 \end{pmatrix} = \begin{pmatrix} \cos 2\theta & \sin 2\theta \\ \sin 2\theta & -\cos 2\theta \end{pmatrix}.$$

E.g., for $\theta = \pi/2$, we obtain $\sigma_\ell = \begin{pmatrix} -1 & 0 \\ 0 & 1 \end{pmatrix} = \sigma_v$ as expected.

EXERCISE 5.8. Verify $\sigma_\ell^2 = \iota$.

5.4 Change of basis. We have associated a matrix to a linear map using the standard basis vectors $E_1 = (1, 0)$ and $E_2 = (0, 1)$. A **basis** of $\mathcal{V}$ is a pair of nonzero and nonparallel vectors, i.e., nonzero vectors X_1, X_2 such that the lines ℓ_i consisting of the points rX_i are not parallel. To a linear map α we can associate a matrix $A = (\alpha_{ij})$ with respect to any basis X_1, X_2 by

$$\alpha(X_j) = \alpha_{1j}X_1 + \alpha_{2j}X_2 \qquad (j = 1, 2). \tag{9}$$

Formulas (5) correspond to the special case where $X_1 = E_1$, $X_2 = E_2$.

Let X_1', X_2' be a further basis and $A' = (\alpha'_{ij})$ the matrix corresponding to the linear map α by

$$\alpha(X_j') = \alpha'_{1j}X_1' + \alpha'_{2j}X_2' \qquad (j = 1, 2). \tag{10}$$

We want to determine the relation between the matrices A and A'.

For this purpose consider the linear map $\gamma: \mathcal{V} \to \mathcal{V}$ defined by $\gamma(X_j) = X_j'$ $(j = 1, 2)$. Let $C = (\gamma_{ij})$ be its matrix with respect to the basis X_1, X_2, i.e.,

$$\gamma(X_j) = X_j' = \gamma_{1j}X_1 + \gamma_{2j}X_2. \tag{11}$$

We claim that

$$AC = CA'. \tag{12}$$

Proof: The following summations are all from 1 to 2. To compute the LHS of (12), we consider

$$\alpha(X_j') = \alpha\gamma(X_j) = \alpha\left(\sum_k \gamma_{kj}X_k\right) = \sum_k \gamma_{kj}\alpha(X_k) = \sum_k \gamma_{kj}\left(\sum_i \alpha_{ik}X_i\right)$$

$$= \sum_i \left(\sum_k \alpha_{ik}\gamma_{kj}\right)X_i = \sum_i (AC)_{ij}X_i.$$

On the other hand

$$\alpha(X'_j) = \sum_k \alpha'_{kj} X'_k = \sum_k \alpha'_{kj} \gamma(X_k) = \sum_k \alpha'_{kj} \left(\sum_i \gamma_{ik} X_i \right) = \sum_i \left(\sum_k \gamma_{ik} \alpha_{kj} \right) X_i$$
$$= \sum_i (CA')_{ij} X_i.$$

By comparison with the previous result we find (for $j = 1, 2$)

$$(AC)_{1j} X_1 + (AC)_{2j} X_2 = (CA')_{1j} X_1 + (CA')_{2j} X_2.$$

Using Proposition 1.18 or Corollary 1.19 we conclude

$$(AC)_{ij} = (CA')_{ij} \qquad (i, j = 1, 2)$$

which proves (12).

Note that the matrix C has an inverse C^{-1}, since the linear map γ has an inverse γ^{-1}. Formula (12) is equivalent to the conjugation formula

$$A = CA'C^{-1}. \tag{13}$$

5.5 The trace of a linear map. First let $A = (\alpha_{ij})$ be a matrix. The *trace of A* is the number

$$\text{trace } A = \alpha_{11} + \alpha_{22}.$$

Let A and B be matrices. We claim that

$$\text{trace}(AB) = \text{trace}(BA). \tag{14}$$

Proof: For the matrix product AB we have by (8),

$$(AB)_{ik} = \sum_j \alpha_{ij} \beta_{jk}$$

and

$$\text{trace}(AB) = \sum_{i,j} \alpha_{ij} \beta_{ji}.$$

Similarly

$$(BA)_{ik} = \sum_j \beta_{ij} \alpha_{jk}$$

and

$$\text{trace}(BA) = \sum_{i,j} \beta_{ij} \alpha_{ji}$$

which proves (14).

The **trace of a linear map** $\alpha : \mathcal{V} \to \mathcal{V}$ is defined by

$$\text{trace } \alpha = \text{trace } A = \alpha_{11} + \alpha_{22}$$

where $A = (\alpha_{ij})$ is the matrix associated to α with respect to a basis X_1, X_2. To show that this number is well-defined, we have to consider the matrix $A' = (\alpha'_{ij})$ associated to α with respect to another basis X'_1, X'_2, and prove that

$$\text{trace } A = \text{trace } A'. \tag{15}$$

Proof: With the notations of Section 5.4 we have $A = CA'C^{-1}$. Using (14) we find

$$\text{trace } A = \text{trace}(CA'C^{-1}) = \text{trace}((A'C^{-1})C) = \text{trace}(A'(C^{-1}C)) = \text{trace } A'$$

5.6 The determinant of a linear map.
Again, first let $A = (\alpha_{ij})$ be a matrix. The **determinant** of A is the number

$$\det A = \begin{vmatrix} \alpha_{11} & \alpha_{12} \\ \alpha_{21} & \alpha_{22} \end{vmatrix} = \alpha_{11}\alpha_{22} - \alpha_{21}\alpha_{12}.$$

Let A and B be matrices. We claim that

$$\det(BA) = \det B \cdot \det A. \tag{16}$$

Proof: The determinant of BA is given by

$$(\beta_{11}\alpha_{11} + \beta_{12}\alpha_{21})(\beta_{21}\alpha_{12} + \beta_{22}\alpha_{22}) - (\beta_{21}\alpha_{11} + \beta_{22}\alpha_{21})(\beta_{11}\alpha_{12} + \beta_{12}\alpha_{22})$$
$$= \beta_{11}\alpha_{11}\beta_{22}\alpha_{22} + \beta_{12}\alpha_{21}\beta_{21}\alpha_{12} - \beta_{21}\alpha_{11}\beta_{12}\alpha_{22} - \beta_{22}\alpha_{21}\beta_{11}\alpha_{12}$$
$$= (\beta_{11}\beta_{22} - \beta_{12}\beta_{21})(\alpha_{11}\alpha_{22} - \alpha_{21}\alpha_{12}) = \det B \cdot \det A.$$

COROLLARY 5.2. *Let C be a matrix with inverse matrix C^{-1}. Then $\det C \neq 0$ and*

$$\det C^{-1} = \frac{1}{\det C}.$$

Proof: By assumption $CC^{-1} = I$ and $\det I = 1$. Therefore the result follows from (16).

The ***determinant of a linear map*** $\alpha : \mathcal{V} \to \mathcal{V}$ is defined by

$$\det \alpha = \det A = \begin{vmatrix} \alpha_{11} & \alpha_{12} \\ \alpha_{21} & \alpha_{22} \end{vmatrix}$$

where $A = (\alpha_{ij})$ is the matrix associated to α with respect to a basis X_1, X_2. To show that this number is well-defined, we have to consider the matrix $A' = (\alpha'_{ij})$ associated to α with respect to another basis X'_1, X'_2, and prove that

$$\det A = \det A'. \tag{17}$$

Proof: With the notations of Section 5.4 we have $A = CA'C^{-1}$. Using (16) we find

$$\det A = \det(CA'C^{-1}) = \det C \cdot \det A' \cdot \det C^{-1} = \det A'.$$

EXERCISE 5.9. Verify that for a rotation ρ_θ we have $\det \rho_\theta = 1$, and for a reflection σ_ℓ in a line ℓ through the origin $\det \sigma_\ell = -1$.

5.7 The crystallographic restriction. A wallpaper pattern has the property that the translational symmetries of the pattern are of the form

$$\mathcal{T} = \{ \tau_A^m \tau_B^n \mid m, n \in \mathbf{Z} \}, \tag{18}$$

where A and B are vectors which form a basis of $\mathcal{V}$. Note that $\mathcal{T}$ is a group and, in particular, is not trivial.

A wallpaper (symmetry) group $\mathcal{W}$ is a group of isometries whose translations are of the form $\mathcal{T}$ above. It is an interesting fact that there are exactly seventeen possible wallpaper groups. We want to explain the key result which leads to the classifications of wallpaper groups. This is the so-called *crystallographic restriction condition.*

Let $\mathcal{W}$ be a wallpaper group and ρ a rotation in $\mathcal{W}$. We state, but do not prove the fact, that ρ is necessarily of finite order. This implies that ρ is a rotation through an angle $\theta = 2\pi/n$, where n is the order of ρ.

THEOREM 5.3 (Crystallographic Restriction). *Let ρ be a rotation in a wallpaper group. Then its order n is either 1, 2, 3, 4 or 6.*

Proof: Let A, B be the vectors occurring in (18), and $\begin{pmatrix} \alpha_{11} & \alpha_{12} \\ \alpha_{21} & \alpha_{22} \end{pmatrix}$ the matrix of ρ with respect to this basis. By Proposition 4.43 we have

$$\rho \tau_A \rho^{-1} = \tau_{\rho(A)}, \; \rho \tau_B \rho^{-1} = \tau_{\rho(B)}.$$

Since $\tau_{\rho(A)} \in \mathcal{T}$, we have

$$\rho(A) = mA + nB = \alpha_{11}A + \alpha_{21}B$$

with integers α_{11} and α_{21}. Similarly

$$\rho(B) = \alpha_{12}A + \alpha_{22}B$$

with integers α_{12} and α_{22}. It follows in particular that the real number trace ρ is an integer.

Next we exploit the fact that trace ρ can also be calculated from the matrix

$$\begin{pmatrix} \cos\theta & -\sin\theta \\ \sin\theta & \cos\theta \end{pmatrix}$$

associated to ρ with respect to the standard basis E_1, E_2. It follows that $2\cos\theta$ is an integer, i.e.,

$$2\cos\theta = 0, \; \pm1, \; \pm2.$$

Thus the only possible cases for n in $\theta = 2\pi/n$ are $n = 1, 2, 3, 4$ or 6.

This is one of the well-established theorems of crystallography: crystals can only have twofold, threefold, fourfold, or sixfold axes of rotational symmetries. No other possibilities, such as fivefold, or sevenfold symmetry, are allowed.

In contrast to crystals, glassy materials have a highly disorganized atomic arrangement. It was a long held view that pure solids are either crystalline or glassy. It seems that this point of view has to be changed. A more recent theory proposes a new phase of solid matter called a quasicrystal. A quasicrystal can have any rotational symmetry. In particular alloys have been discovered with a fivefold symmetry axis.

A few years earlier, Penrose had discovered curious mosaic patterns which could be fitted together to fill a plane. His goal had nothing to do with atomic structures, but with nonperiodic tilings of the plane. These Penrose patterns and their properties have given rise to a new theory of quasicrystals. Even plane geometry is still a source of surprises and evolving.

A ANSWERS TO ODD-NUMBERED EXERCISES

Chapter 1.

1.1.
$$r((a_1, a_2) + (b_1, b_2)) = r(a_1 + b_1, a_2 + b_2)$$
$$= (r(a_1 + b_1), r(a_2 + b_2))$$
$$= (ra_1 + rb_1, ra_2 + rb_2)$$
$$= (ra_1, ra_2) + (rb_1, rb_2) = rA + rB.$$

1.3. $O = r(O + O) = rO + rO$, which implies by Exercise 1.2 that $O = rO$.

1.5. Assume $r \neq 0$. Then

$$\frac{1}{r}(rA) = \left(\frac{1}{r} \cdot r\right) A = 1A = A.$$

But $rA = O$ by assumption, so the left hand side is O by Exercise 1.3, hence $A = O$.

1.7. (a) $M_1 = \frac{1}{2}(A + B)$, $M_2 = \frac{1}{2}(B + C)$, $M_3 = \frac{1}{2}(C + D)$, $M_4 = \frac{1}{2}(D + A)$.

$M_1 + M_3$ and $M_2 + M_4$ both give the same expression $\frac{1}{2}(A+B+C+D)$, which proves the desired statement by formula (13).

(b) For $A = B$ the statement is that in a triangle a certain parallelogram figure appears (make a drawing). For $A = B = C$ the parallelogram in question degenerates to a line segment and its midpoint.

1.9. The three diagonals have a common intersection point.

1.11. The triangle $\triangle A'B'C'$ of midpoints has as centroid

$$G' = \frac{1}{3}(A' + B' + C') = \frac{1}{6}(B + C + A + C + A + B) = G.$$

1.13. Let X be the centroid of $\triangle ABC$. Thus $X = \frac{1}{3}(A+B+C)$. Then $X - B = \frac{1}{3}(A + C) - \frac{2}{3}B = \frac{1}{3}(B + D) - \frac{2}{3}B = \frac{1}{3}(D - B)$. Similarly, for the centroid Y

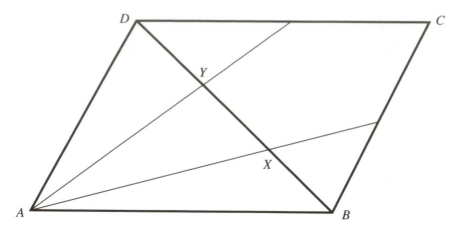

Figure A.1.

of $\triangle ACD$, $Y - D = \frac{1}{3}(B - D)$. These formulas prove the stated result. The situation is illustrated in Figure A.1. An alternate argument is as follows. The lines ℓ_{AX} and ℓ_{BD} are medians of $\triangle ABC$, hence intersect in X. Thus

$$ X - B = \frac{2}{3} \cdot \frac{1}{2}(D - B) = \frac{1}{3}(D - B). $$

Similarly for Y.

1.15. (a) The quadrilateral and the six possible midpoints look as in Figure A.2.

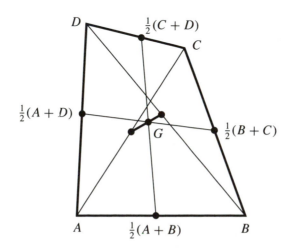

Figure A.2.

Then

$$\frac{1}{2}\left[\frac{1}{2}(A+D)+\frac{1}{2}(B+C)\right] = \frac{1}{2}\left[\frac{1}{2}(A+B)+\frac{1}{2}(C+D)\right]$$

$$= \frac{1}{2}\left[\frac{1}{2}(A+C)+\frac{1}{2}(B+D)\right]$$

$$= \frac{1}{4}(A+B+C+D) = G.$$

If one thinks of A, B, C, D as the vertices of a tetrahedron (projected to a plane), then the three lines in question join the midpoints of opposite edges, and intersect in their common midpoint, which is the centroid of A, B, C, D.

(b) The geometric meaning can be read off from Figure A.3.

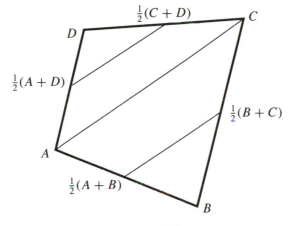

Figure A.3.

1.17. This follows from Figure 1.11 and the construction of the barycentric coordinates in the proof of Theorem 1.11. The points with constant $a = x$ are those projecting to the same X on ℓ_{CA}, i.e., the points of a line parallel to ℓ_{BC}. An alternative proof would be the following kinematic argument. The points with fixed barycentric coordinate a are of the form $P(t) = aA+tB+(1-a-t)C$, where $t = b$ is thought of as time parameter. The velocity of $P(t)$ is then given by the derivative $\dot{P}(t) = \frac{d}{dt}P(t) = B - C$. But this vector is parallel to the line ℓ_{BC}, so that $P(t)$ moves on a line parallel to ℓ_{BC}.

1.19. It suffices to consider the case of a zero ζ of f' which is distinct from the z_i ($i = 1, \ldots, n$). Then, as in the proof of Theorem 1.12

$$\frac{f'(\zeta)}{f(\zeta)} = \sum_{i=1}^{n} \frac{1}{\zeta - z_i} = 0,$$

which yields

$$\sum_{i=1}^{n} \frac{\zeta - z_i}{|\zeta - z_i|^2} = 0,$$

and the desired representation $\zeta = \sum_{i=1}^{n} a_i z_i$, with

$$a_i = \frac{1}{\sum_{j=1}^{n} \dfrac{1}{|\zeta - z_j|^2}} \cdot \frac{1}{|\zeta - z_i|^2} \qquad (i = 1, \ldots, n).$$

1.21. Let a, b, c be the barycentric coordinates of G, i.e.,

$$G = aA + bB + cC \qquad \text{with } a + b + c = 1.$$

By the proof of Theorem 1.13, we have $A' = \dfrac{bB + cC}{b + c}$ and $G = aA + (b + c)A'$, which by (21), (23) yields

$$\frac{G - A}{G - A'} = -\frac{b + c}{a}.$$

Similarly

$$\frac{G - B}{G - B'} = -\frac{a + c}{b}, \qquad \frac{G - C}{G - C'} = -\frac{a + b}{c}.$$

Thus

$$\frac{G - A}{G - A'} + \frac{G - B}{G - B'} + \frac{G - C}{G - C'} = -\left(\frac{b + c}{a} + \frac{a + c}{b} + \frac{a + b}{c}\right)$$

$$= -\left[\left(\frac{b}{a} + \frac{a}{b}\right) + \left(\frac{c}{b} + \frac{b}{c}\right) + \left(\frac{a}{c} + \frac{c}{a}\right)\right].$$

For every $x > 0$, $2 \leq x + \dfrac{1}{x}$ with equality holding if and only if $x = 1$. It follows that for points inside $\triangle ABC$ (hence $a > 0$, $b > 0$, $c > 0$) we have

$$\frac{G - A}{G - A'} + \frac{G - B}{G - B'} + \frac{G - C}{G - C'} \leq -6,$$

with equality holding if and only if $\dfrac{b}{a} = 1$, $\dfrac{c}{b} = 1$, $\dfrac{a}{c} = 1$, i.e., if and only if $a = b = c = \dfrac{1}{3}$. This is the case precisely when G is the centroid of $\triangle ABC$.

1.23. With the notations as in Theorem 1.13, consider $\triangle AC'C$. Apply Menelaus to the transversal $\ell_{BB'}$ (intersecting all sides of $\triangle AC'C$). Then

$$\frac{G - C'}{G - C} \cdot \frac{B' - C}{B' - A} \cdot \frac{B - A}{B - C'} = 1.$$

Similarly $\ell_{AA'}$ is a transversal to $\triangle C'BC$, and thus

$$\frac{A' - B}{A' - C} \cdot \frac{G - C}{G - C'} \cdot \frac{A - C'}{A - B} = 1.$$

Multiplying these expressions yields the identity

$$\frac{B' - C}{B' - A} \cdot \frac{C' - A}{C' - B} \cdot \frac{A' - B}{A' - C} = -1$$

which is the implication in Ceva's Theorem. The converse is proved as in the original proof.

Chapter 2.

2.1. The composition of one-to-one maps is one-to-one. The composition of maps which are both onto is an onto map. Thus the composition of bijections is a bijection.

2.3. $\dfrac{1}{r}P = \dfrac{aA + bB}{r}$ with $\dfrac{a}{r} + \dfrac{b}{r} = 1$. Thus $\dfrac{1}{r}P \in \ell_{AB}$ and $P \in \delta_r(\ell_{AB})$, a line parallel to ℓ_{AB}.

2.5. There is a unique r such that $\delta_{D,r}(A) = A'$. The assumption implies that $\delta_{D,r}(\ell_{AB}) = \ell_{A'B'}$. It follows that $\delta_{D,r}(B) = B'$. Similarly $\delta_{D,r}(C) = C'$. Therefore $\delta_{D,r}(\ell_{AC}) = \ell_{A'C'}$, which proves $\ell_{AC} \parallel \ell_{A'C'}$.

2.7. Apply the dilatation $\delta_{G,-2}$, which maps $\triangle A'B'C'$ into $\triangle ABC$, and thus $\ell_{A'P}$ to line a, $\ell_{B'P}$ to b, and $\ell_{C'P}$ to c. The argument is now as in the proof of Theorem 2.12.

2.9. The necessary condition for $\tau_A = \sigma_Q \sigma_P$ is $A = 2(Q - P)$. For prescribed Q there is a unique $P = Q - \frac{1}{2}A$ satisfying this condition.

2.11. We calculate

$$(\tau_A \sigma_P \tau_A^{-1})(X) = (\tau_A \sigma_P)(X - A) = (2P - (X - A)) + A = 2(P + A) - X,$$

which equals $\sigma_Q(X)$ for $Q = P + A$ or $A = Q - P$.

2.13. The commutativity condition assumed implies by (33) the identity

$$s(1 - r)A + (1 - s)B = r(1 - s)B + (1 - r)A.$$

Collecting terms we have

$$(s - 1)(1 - r)A + (1 - s)(1 - r)B = O$$

or

$$(1 - s)(1 - r)(B - A) = O.$$

Since $r \neq 1$ and $s \neq 1$, this implies $B - A = O$.

2.15. $E = EE' = E'$.

2.17. Note that

$$X^{-1}(X^{-1})^{-1} = E = (X^{-1})^{-1}X$$
$$X^{-1}X = E = XX^{-1},$$

so $(X^{-1})^{-1} = X$ by Exercise 2.16.

2.19. First we prove (37). Fix $n \in \mathbf{Z}$. We make an induction argument on m. Consider the case $m > 0$. If $m = 1$, then (37) claims that $X \cdot X^n = X^{n+1}$. This is certainly true for $n \geq 0$. If $n < 0$, then

$$X \cdot (X^{-1})^{-n} = X \cdot (X^{-1}(X^{-1})^{-n-1}) = (XX^{-1})(X^{-1})^{-n-1} = (X^{-1})^{-(n+1)}$$

which equals X^{n+1}, so (37) holds for $m = 1$ and all $n \in \mathbf{Z}$. If (37) holds for an $m > 0$, it also holds for $m + 1$, since

$$X^{m+1} \cdot X^n = X \cdot X^m \cdot X^n = X \cdot X^{m+n} = X^{(m+1)+n}.$$

The case $m = 0$ amounts to the formula $EX^n = X^{0+n} = X^n$. Finally consider the case $m < 0$. Then

$$X^m X^n = (X^{-1})^{-m}X^n = (X^{-1})^{-m}(X^{-1})^{-n} = (X^{-1})^{-m-n}$$

since $-m > 0$. But the last term equals X^{m+n}.

Now we prove (38). Fix $m \in \mathbf{Z}$. We make an induction argument on n. Consider the case $n > 0$. If $n = 1$, then (38) claims that $(X^m)^1 = X^m$. The induction step from n to $n + 1$ is as follows:

$$(X^m)^{n+1} = (X^m)^n X^m = X^{m \cdot n} X^m = X^{mn+m} = X^{m(n+1)}.$$

If $n = 0$, (38) amounts to $(X^m)^0 = E = X^{m \cdot 0}$. We now have (38) for all m and all $n \geq 0$. Finally assume $n < 0$. Note that $(X^m)^{-1} = X^{-m}$, since $X^m \cdot X^{-m} = X^{-m} \cdot X^m = X^0 = E$. Then

$$(X^m)^n = ((X^m)^{-1})^{-n} = (X^{-m})^{-n} \qquad \text{by (37).}$$

This equals $X^{(-m)(-n)} = X^{mn}$ by the case considered before.

2.21. An element $k \in \mathbf{Z}_n$ is the multiple $k \cdot 1$ of 1.

2.23. They are the same.

2.25. If $\mathcal{G} = \{E, X, Y\}$, then the multiplication table is necessarily

	E	X	Y
E	E	X	Y
X	X	Y	E
Y	Y	E	X

Thus $h(E) = 0$, $h(X) = 1$, $h(Y) = 2$ provides an isomorphism of $\mathcal{G}$ with $\mathbf{Z}_3$.

2.27. Let $\mathcal{G}$ be a group with four elements. If there exists an element $X \in \mathcal{G}$ of order 4, then $h: \mathcal{G} \to \mathbf{Z}_4$ given by $h(E) = 0$, $h(X) = 1$, $h(X^2) = 2$, $h(X^3) = 3$ defines an isomorphism of $\mathcal{G}$ with $\mathbf{Z}_4$. Assume now that $\mathcal{G}$ has no element of order 4. It remains to show that there is an isomorphism of $\mathcal{G}$ with $\mathcal{V}_4$. The essential point is that in $\mathcal{V}_4$ every element different from the identity is an involution. Thus it is sufficient to show that every element $X \neq E$ in $\mathcal{G}$ has order two (provided $\mathcal{G}$ is not isomorphic with $\mathbf{Z}_4$). Assume to the contrary that $X^2 = Y \neq E$. If one knows the fact that the order of any element of $\mathcal{G}$ divides the order of $\mathcal{G}$, one can conclude that X is of order four and therefore $\mathcal{G} = \langle X \rangle$ is isomorphic to $\mathbf{Z}_4$, a contradiction. If one does not want to use this fact, one can argue from Proposition 2.21 that the multiplication table of $\mathcal{G}$ is necessarily

	E	X	Y	Z
E	E	X	Y	Z
X	X	Y	Z	E
Y	Y	Z	E	X
Z	Z	E	X	Y

Thus $k(E) = 0$, $k(X) = 1$, $k(Y) = 2$, $k(Z) = 3$ defines an isomorphism of $\mathcal{G}$ with $\mathbf{Z}_4$, again a contradiction.

2.29. $\mathcal{D}_{20}$ (ignoring the bolt holes, otherwise $\mathcal{D}_5$).

Chapter 3.

3.1. E.g., (SP2) for $X = (x_1, x_2)$, $Y = (y_1, y_2)$ and $Z = (z_1, z_2)$ amounts to the identity

$$(x_1 + y_1)z_1 + (x_2 + y_2)z_2 = (x_1 z_1 + x_2 z_2) + (y_1 z_1 + y_2 z_2).$$

3.3. With $C = O$ we have $A' = 1/2 \cdot B$ and $B' = 1/2 \cdot A$. The assumption is that $|A' - A| = |B' - B|$. The conclusion to reach is that $|A| = |B|$. But

$$|A' - A| = \left| \frac{1}{2}B - A \right| \quad \text{and} \quad |B' - B| = \left| \frac{1}{2}A - B \right|$$

implies

$$|A' - A|^2 = \frac{1}{4}|B|^2 - B \cdot A + |A|^2$$

$$|B' - B|^2 = \frac{1}{4}|A|^2 - A \cdot B + |B|^2.$$

The hypothesis therefore implies

$$|A|^2 + \frac{1}{4}|B|^2 = |B|^2 + \frac{1}{4}|A|^2$$

or

$$\frac{3}{4}|A|^2 = \frac{3}{4}|B|^2.$$

Hence $|A| = |B|$.

3.5. Assume $O \in \ell_{AB}$, i.e., $aA + bB = O$ for some real numbers with $a + b = 1$. If say $a \neq 0$, then $A = -\frac{b}{a}B$ and $A \cdot B = -\frac{b}{a}|B|^2 = 0$, hence $b = 0$. Therefore $aA = O$ and thus $A = O$, a contradiction.

3.7. (i) $A' - B' = \frac{1}{2}(B - A)$ and $B'' - A'' = \frac{1}{2}(B - A)$, which shows $A' - B' = B'' - A''$, and proves that $A''B''A'B'$ is a parallelogram.

(ii) $\ell_{A'B'}$ is parallel to ℓ_{AB}, by what we just proved. Using the same argument in $\triangle BCH$, it follows that $\ell_{A'B''}$ is parallel to the altitude ℓ_C, hence perpendicular to ℓ_{AB}, which proves that $A''B''A'B'$ is a rectangle.

3.9. By definition $|X - A|^2 = r^2|X - B|^2$ or

$$|X|^2 - 2X \cdot A + |A|^2 = r^2(|X|^2 - 2X \cdot B + |B|^2),$$

hence

$$(r^2 - 1)|X|^2 + 2X \cdot (A - r^2 B) + r^2|B|^2 - |A|^2 = 0,$$

hence

$$|X|^2 + 2X \cdot \frac{A - r^2 B}{r^2 - 1} = \frac{|A|^2 - r^2 |B|^2}{r^2 - 1}.$$

Complete the square to obtain

$$\left| X + \frac{A - r^2 B}{r^2 - 1} \right|^2 = \frac{|A|^2 - r^2 |B|^2}{r^2 - 1} + \left| \frac{A - r^2 B}{r^2 - 1} \right|^2.$$

The RHS equals $\dfrac{r^2}{(r^2 - 1)^2} |A - B|^2$, so that $\mathcal{C}_r$ is a circle with center $\dfrac{r^2 B - A}{r^2 - 1}$ and radius $\dfrac{r}{|r^2 - 1|} \cdot |A - B|$.

To get an intuitive feel for this, consider the points $X = C$, $X = D$ to be the two points on ℓ_{AB} such that

$$\left| \frac{C - A}{C - B} \right| = r \quad \text{and} \quad \left| \frac{D - A}{D - B} \right| = r.$$

If we know that $\mathcal{C}_r$ is a circle, its center is the midpoint of CD and its radius $\frac{1}{2}|C - D|$.

If the origin is in particular the midpoint of AB, then $B = -A$. It follows that

$$C - A = r(C + A),$$

or

$$(1 - r)C = (1 + r)A,$$

which implies

$$C = \frac{1 + r}{1 - r} \cdot A$$

Similarly,

$$D - A = -r(D + A)$$

implies

$$(1 + r)D = (1 - r)A,$$

or

$$D = \frac{1 - r}{1 + r} A.$$

Thus the center of the circle $\mathcal{C}_r$ must be

$$\frac{1}{2}(C + D) = \frac{1}{2} \left[\frac{1 + r}{1 - r} + \frac{1 - r}{1 + r} \right] A = \frac{1 + r^2}{1 - r^2} \cdot A.$$

The formula given above yields for $B = -A$ the same expression.

For the radius we obtain

$$\frac{1}{2}|C - D| = \frac{1}{2} \left| \frac{1 + r}{1 - r} - \frac{1 - r}{1 + r} \right| \cdot |A| = \frac{2r}{|1 - r^2|} \cdot |A|.$$

The formula given above yields for $B = -A$ the same expression.

3.11.

$$d(X, Y) + d(Y, Z) = |X - Y| + |Y - Z| \geq |(X - Y) + (Y - Z)|$$
$$= |X - Z| = d(X, Z).$$

3.13. X is the sum of its projections to the orthogonal lines spanned by E_1 and E_2. Those projections are $(X \cdot E_i)E_i$ for $i = 1, 2$.

3.15. The height of $\triangle OXY$ is

$$h = |Y - \text{proj}_X Y| = |Y| |\sin \angle (X, Y)|.$$

Thus

$$\text{area}(\triangle OXY) = \frac{1}{2} \cdot |X| |Y| |\sin \angle (X, Y)|.$$

Alternatively, the formula

$$\cos \angle (X, Y) = \frac{X \cdot Y}{|X||Y|}$$

implies

$$\sin^2 \angle (X, Y) = 1 - \frac{(X \cdot Y)^2}{|X|^2 |Y|^2}.$$

Then by Exercise 3.14,

$$\text{area}(\triangle OXY) = \frac{1}{2} \cdot \sqrt{|X|^2 |Y|^2 - (X \cdot Y)^2}$$

$$= \frac{1}{2} |X| |Y| |\sin \angle (X, Y)|.$$

3.17. (26) reads

$$x_1 n_1 + x_2 n_2 = p_1 n_1 + p_2 n_2,$$

or in familiar form,

$$ax_1 + bx_2 + c = 0$$

with $a = n_1$, $b = n_2$, $c = -(p_1 n_1 + p_2 n_2)$.

Chapter 4.

4.1. If α is a translation, this is clear. Thus we can assume that $\alpha(O) = O$. We have then for vectors X, Y

$$\cos \angle (X, Y) = \frac{XY}{|X||Y|}, \qquad \cos \angle (\alpha(X), \alpha(Y)) = \frac{\alpha(X) \cdot \alpha(Y)}{|\alpha(X)| |\alpha(Y)|}.$$

Since α is an isometry, we have

$$|\alpha(X)| = |X|, \quad |\alpha(Y)| = |Y|, \quad \alpha(X) \cdot \alpha(Y) = X \cdot Y.$$

It follows that $\cos \angle (X, Y) = \cos \angle (\alpha(X), \alpha(Y))$ and $\angle (X, Y) = \pm \angle (\alpha(X), \alpha(Y))$.

4.3. Clearly $\sigma_\ell(O) = O$. By the first part of the proof of Theorem 4.3, it suffices to show that $|\sigma_\ell(X)| = |X|$ for all vectors X.

Consider the quadrilateral O, $\sigma_\ell(X)$, $\sigma_\ell(X) + X$, X. It is clearly a parallelogram. We have to show that it is a rhombus. This is equivalent to the statement that the diagonal vectors $\sigma_\ell(X) - X$ and $\sigma_\ell(X) + X$ are orthogonal. In terms of a unit vector Y on ℓ we have by (4),

$$\sigma_\ell(X) + X = 2(X \cdot Y)Y.$$

Further, by (6),

$$\sigma_\ell(X) - X = 2U,$$

where $U = (X \cdot Y)Y - X$. It follows that

$$(\sigma_\ell(X) - X) \cdot (\sigma_\ell(X) + X)$$

equals $4X \cdot Y$ times

$$((X \cdot Y)Y - X) \cdot Y = (X \cdot Y)(Y \cdot Y) - X \cdot Y = 0,$$

since Y is a unit vector. This completes the proof.

4.5. The equality $\alpha\sigma_\ell\alpha^{-1} = \sigma_{\alpha(\ell)}$ proves that under the stated assumption $\sigma_{\alpha(\ell)} = \sigma_\ell$ for all lines ℓ. Since the fixed points of a reflection in a line are the points of that line, this shows that $\alpha(\ell) = \ell$ for all lines. Given any point P, consider two lines ℓ and ℓ', through P. Since $\alpha(\ell) = \ell$ and $\alpha(\ell') = \ell'$, it follows that $\alpha(P) = P$. Thus α must not only map every line to itself, but also leave every point fixed. Thus $\alpha = \iota$.

4.7. Let m and n be perpendicular lines intersecting in P. One proof that $\sigma_n\sigma_m = \sigma_P$ follows directly from the geometric definition of the maps involved.

Since n and m are perpendicular, $\sigma_m(n) = n$, so by Proposition 4.13 implies that $\sigma_m\sigma_n\sigma_m = \sigma_n$. Thus

$$\sigma_n\sigma_m\sigma_n\sigma_m = (\sigma_n\sigma_m)^2 = \iota.$$

Since $\sigma_n\sigma_m = \iota$ would imply $\sigma_m = \sigma_n$, it follows that $\sigma_n\sigma_m$ is an involution. Moreover, $P = m \cap n$ implies

$$(\sigma_n\sigma_m)(P) = \sigma_n(\sigma_m(P)) = \sigma_n(P) = P,$$

so P is a fixed point of $\sigma_n\sigma_m$. To show that $\sigma_n\sigma_m = \sigma_P$, it suffices by Theorem 4.15 to show that no other point X is a fixed point of $\sigma_n\sigma_m$. But $(\sigma_n\sigma_m)(X) = X$ implies $\sigma_m(X) = \sigma_n(X)$, which is impossible unless $X = P$.

If m and n intersect in the origin, a proof using formula (4) goes as follows. Let M be a unit vector on m and N a unit vector on n. Then

$$\sigma_m(X) = -X + 2(X \cdot M)M$$

and

$$
\begin{aligned}
(\sigma_n \sigma_m)(X) &= -\sigma_m(X) + 2(\sigma_m(X) \cdot N)N \\
&= X - 2(X \cdot M)M + 2((-X + 2(X \cdot M)M) \cdot N)N \\
&= X - 2(X \cdot M)M - 2(X \cdot N)N + 4(X \cdot M)(M \cdot N)N.
\end{aligned}
$$

Since M and N are orthogonal, and

$$X = (X \cdot M)M + (X \cdot N)N,$$

it follows that

$$(\sigma_n \sigma_m)(X) = -X$$

and $\sigma_n \sigma_m$ is indeed the central reflection in the origin.

The general case of lines m and n intersecting in an arbitrary point P can be reduced to this special case by consideration of the reflections

$$\sigma_{\tau_P^{-1}(m)} = \tau_P^{-1} \sigma_m \tau_P \quad \text{and} \quad \sigma_{\tau_P^{-1}(n)} = \tau_P^{-1} \sigma_n \tau_P.$$

The lines $\tau_P^{-1}(m)$ and $\tau_P^{-1}(n)$ are orthogonal and intersect in the origin. Therefore by the special case discussed,

$$\sigma_{\tau_P^{-1}(n)} \sigma_{\tau_P^{-1}(m)} = -\iota.$$

It follows that

$$\tau_P^{-1} \sigma_n \tau_P \tau_P^{-1} \sigma_m \tau_P = -\iota = \sigma_O.$$

Thus

$$\sigma_n \sigma_m = \tau_P \sigma_O \tau_P^{-1} = \sigma_P.$$

4.9. The identity $\sigma_\ell \sigma_B \sigma_\ell \sigma_A = \iota$ is equivalent to

$$\sigma_{\sigma_\ell(B)} = \sigma_A,$$

which in turn is equivalent to $\sigma_\ell(B) = A$. This is the case precisely when ℓ is the perpendicular bisector of A and B.

4.11. With the notations as in Figure 4.4, we have the following equivalent properties: $\sigma_n \sigma_m = \sigma_m \sigma_n \iff \tau_{2(N-M)} = \tau_{2(M-N)} \iff N - M = M - N \iff N = M \iff n = m$.

4.13. Let M be any point, and denote $N = M + 1/2 \cdot A$. Then $\tau_A = \sigma_N \sigma_M$. It follows that

$$\alpha \tau_A \alpha^{-1} = \alpha \sigma_N \sigma_M \alpha^{-1} = (\alpha \sigma_N \alpha^{-1})(\alpha \sigma_M \alpha^{-1})$$

$$= \sigma_{\alpha(N)} \sigma_{\alpha(M)}$$

$$= \tau_{2(\alpha(N) - \alpha(M))}.$$

For a linear isometry α we have

$$2(\alpha(N) - \alpha(M)) = 2\alpha(N - M) = 2\alpha(A/2) = \alpha(A),$$

and $\alpha \tau_A \alpha^{-1} = \tau_{\alpha(A)}$.

4.15. For any point G

$$(\sigma_G \sigma_{A_n})(\sigma_G \sigma_{A_{n-1}}) \cdots (\sigma_G \sigma_{A_1}) = \tau_{2[nG - (A_1 + A_2 + \cdots + A_n)]}.$$

This map equals the identity if and only if

$$G = \frac{1}{n}(A_1 + A_2 + \cdots + A_n).$$

4.17. By formula (15), with A, B, C counter-clockwise, and α, β, γ positive angles at A, B, C, we have

$$\rho_{B,2\beta} \rho_{C,2\gamma} = \rho_{A,2(\beta+\gamma)}.$$

It follows that

$$\rho_{A,2\alpha} \rho_{B,2\beta} \rho_{C,2\gamma} = \rho_{A,2\pi} = \iota.$$

Similarly,

$$\rho_{C,2\alpha} \rho_{A,2\alpha} \rho_{B,2\beta} = \iota \quad \text{and} \quad \rho_{B,2\beta} \rho_{C,2\gamma} \rho_{A,2\alpha} = \iota.$$

4.19. Let $\rho_{A,\alpha}$ and $\rho_{B,\beta}$ be nontrivial rotations, and assume

$$\rho_{A,\alpha} \rho_{B,\beta} = \rho_{B,\beta} \rho_{A,\alpha}.$$

Since B is a fixed point of $\rho_{B,\beta}$, it follows that

$$\rho_{A,\alpha}(B) = \rho_{B,\beta}(\rho_{A,\alpha}(B)).$$

But B is the only fixed point of $\rho_{B,\beta}$, hence

$$\rho_{A,\alpha}(B) = B.$$

It follows that $A = B$.

4.21. Denote the sides opposite A, B and C by a, b and c, respectively. Then by formula (14)

$$\rho_{C,2\gamma} = \sigma_a\sigma_b, \quad \rho_{B,2\beta} = \sigma_c\sigma_a, \quad \rho_{A,2\alpha} = \sigma_b\sigma_c.$$

It follows that

$$\rho_{C,2\gamma}\rho_{B,2\beta}\rho_{A,2\alpha} = (\sigma_a\sigma_b\sigma_c)^2.$$

Since a, b, c are neither parallel nor concurrent, by Theorem 4.37 the product $\sigma_a\sigma_b\sigma_c$ is a glide reflection. By formula (17) the square $(\sigma_a\sigma_b\sigma_c)^2$ is a nontrivial translation.

Chapter 5.

5.1. Let π_ℓ denote the orthogonal projection of the plane onto ℓ. Then by equation (21) of Chapter 3 we have

$$\pi_\ell(X) = (X \cdot A)A.$$

Thus for $A = a_1E_1 + a_2E_2$ we obtain

$$\pi_\ell(E_1) = (E_1 \cdot A)A = a_1A = a_1^2E_1 + a_1a_2E_2$$
$$\pi_\ell(E_2) = (E_2 \cdot A)A = a_2A = a_2a_1E_1 + a_2^2E_2.$$

It follows that π_ℓ is represented by the matrix

$$\begin{pmatrix} a_1^2 & a_1a_2 \\ a_1a_2 & a_2^2 \end{pmatrix}.$$

If θ denotes the angle between the (positive) x-axis and the line ℓ, or more precisely $\theta = \sphericalangle(E_1, A)$, then $a_1 = \cos\theta$ and $a_2 = \sin\theta$. It follows that the matrix of π_ℓ is

$$\begin{pmatrix} \cos^2\theta & \cos\theta\sin\theta \\ \cos\theta\sin\theta & \sin^2\theta \end{pmatrix}.$$

5.3. This follows from formula (8).

5.5. The result in both cases is the identity matrix. This corresponds to the fact that the maps σ_h and σ_v are both involutions.

5.7. Using (8) and the trigonometric identities below, we find

$$\begin{pmatrix} \cos\theta_2 & -\sin\theta_2 \\ \sin\theta_2 & \cos\theta_2 \end{pmatrix}\begin{pmatrix} \cos\theta_1 & -\sin\theta_1 \\ \sin\theta_1 & \cos\theta_1 \end{pmatrix} = \begin{pmatrix} \cos(\theta_1+\theta_2) & -\sin(\theta_1+\theta_2) \\ \sin(\theta_1+\theta_2) & \cos(\theta_1+\theta_2) \end{pmatrix}.$$

Note that we know a priori that this must be the result, namely the matrix associated to $\rho_{\theta_1+\theta_2}$. Therefore the trigonometric identities

$$\cos(\theta_1 + \theta_2) = \cos\theta_1\cos\theta_2 - \sin\theta_1\sin\theta_2$$
$$\sin(\theta_1 + \theta_2) = \sin\theta_1\cos\theta_2 + \sin\theta_2\cos\theta_1$$

are conversely proved through this matrix multiplication.

5.9. This follows from the matrix representations

$$\rho_\theta = \begin{pmatrix} \cos\theta & -\sin\theta \\ \sin\theta & \cos\theta \end{pmatrix} \qquad \sigma_\ell = \begin{pmatrix} \cos 2\theta & \sin 2\theta \\ \sin 2\theta & -\cos 2\theta \end{pmatrix}$$

and the trigonometric identity $\cos^2\varphi + \sin^2\varphi = 1$, applied to $\varphi = \theta$ and $\varphi = 2\theta$.

B BIBLIOGRAPHY

1. Abelson, H. and diSessa, A., Turtle Geometry, The Computer as a Medium for Exploring Mathematics, M.I.T. Press, Cambridge, MA, 1981.

2. Alperin, J. L., *Groups and Symmetry*, in Mathematics Today, edited by L. A. Steen, Springer-Verlag, New York, 1978.

3. Benson, C. T. and Grove, L. C., Finite Reflection Groups, Bogdon & Quigley, Tarrytown-on-Hudson, NY, 1971.

4. Burn, R. P., Deductive Transformation Geometry, Cambridge University Press, Cambridge, 1975.

5. Coxeter, H. S. M., Introduction to Geometry, Wiley, New York, 1961.

6. Dodge, C. W., Euclidean Geometry and Transformations, Addison-Wesley, Reading, MA, 1972.

7. Eccles, F., An Introduction to Transformational Geometry, Addison-Wesley, Reading, MA, 1971.

8. Ewald, G., Geometry, An Introduction, Wadsworth, Belmont, CA, 1971.

9. Faber, R. L., Foundations of Euclidean and non-Euclidean Geometry, Marcel Dekker, New York, 1983.

10. Greenberg, M. J., Euclidean and non-Euclidean Geometries, Development and History, 2nd ed., Freeman, San Francisco, 1980.

11. Guggenheimer, H. W., Plane Geometry and Its Groups, Holden-Day, San Francisco, 1967.

12. Heath, T. L., The Thirteen Books of Euclid's Elements, 2nd ed., 3 vols., Dover, New York, 1956.

13. Jeger, M., Transformation Geometry, American Elsevier Publishing Company, Inc., New York, 1969.

14. Klein, F., Elementary Mathematics from an Advanced Standpoint (Vol. 2, Geometry), Macmillan, New York, 1939.

15. Lyubich, Yu. I. and Shor, L. A., The Kinematic Method in Geometrical Problems, MIR Publishers, Moscow, 1980.

16. Martin, G. E., Transformation Geometry, An Introduction to Symmetry, Springer-Verlag, New York, 1982.

17. Maxwell, E. A., Geometry by Transformations, Cambridge University Press, Cambridge, 1975.

18. Millman, R. S. and Parker, G. D., Geometry, A Metric Approach with Models, Springer-Verlag, New York, 1981.

19. Moise, E. E., Elementary Geometry from an Advanced Standpoint, 3rd edition, Addison-Wesley, Indianapolis, 1990.

20. Penrose, R., *The Geometry of the Universe*, in Mathematics Today, edited by L. A. Steen, Springer-Verlag, New York, 1978.

21. Ryan, P. J., Euclidean Geometry and Non-Euclidean Geometry, Cambridge University Press, Cambridge, 1986.

22. Weyl, H., Symmetry, Princeton University Press, Prineton, 1952.

23. Yaglom, I. M., Geometric Transformations, New Mathematical Library, Random House, 1962.

24. Yaglom, I. M., Felix Klein and Sophus Lie, Evolution of the Idea of Symmetry in the Nineteenth Century, Birkhäuser, Boston, Basel, 1988.

25. Yale, P. B., Geometry and Symmetry, Holden-Day, San Francisco, 1968.

I INDEX